Know-It-All

Great Inventions

Know-It-All

Great Inventions

**The 50 Greatest Inventions,
Each Explained in Under a Minute >**

Editor **David Boyle**

Contributors

**David Boyle
Judith Hodge
Diana Rawlinson
Andrew Simms**

—WELLFLEET—
P R E S S

Inspiring | Educating | Creating | Entertaining

© 2017 Quarto Publishing plc

First published in 2017 by Wellfleet Press, an imprint of The Quarto Group, 142 West 36th Street, 4th Floor, New York, New York 10018, USA
T (212) 779-4972 **F** (212) 779-6058
www.QuartoKnows.com

Titles are also available at discount for retail, wholesale, promotional, and bulk purchase. For details, contact the Special Sales Manager by email at specialsales@ quarto.com or by mail at The Quarto Group, Attn: Special Sales Manager, 401 Second Avenue North, Suite 310, Minneapolis, MN 55401, USA.

10 9 8 7 6 5 4 3 2 1

ISBN: 978-1-57715-160-9

This book was conceived, designed, and produced by
Ivy Press
An imprint of The Quarto Group
The Old Brewery, 6 Blundell Street
London N7 9BH, United Kingdom
T (0)20 7700 6700 **F** (0)20 7700 8066

Publisher **Susan Kelly**
Creative Director **Michael Whitehead**
Editorial Director **Tom Kitch**
Art Director **James Lawrence**
Project Editors **Fleur Jones, Stephanie Evans**
Designer **Ginny Zeal**
Illustrator **Steve Rawlings**
Assistant Editor **Jenny Campbell**

Printed in China

FSC
www.fsc.org
MIX
Paper from
responsible sources
FSC® C001701

CONTENTS

INTRODUCTION
David Boyle

What makes a successful inventor? There are many ingredients. You need to have the background knowledge in your field, like the Wright Brothers, pioneers of the airplane. You need to know the right people who can help you exploit your invention, as Scottish engineer James Watt did when he famously got together with Birmingham manufacturer Matthew Boulton to make the most of Watt's improvements to the steam engine. Then you need people to understand the significance of what you have created. Without this an invention can lie ignored and unexploited for decades, as Scottish inventor James Blyth found with his wind energy machine. You also need to have investors and you need to be able to protect your invention, as the rival inventors of the pneumatic tire, Robert Thomson and John Boyd Dunlop, discovered. But most of all, you need a moment of inspiration.

Sometimes this is a revelation that is then worked out carefully in theory, as with Archimedes' screw, an ancient Greek invention. Sometimes it is a series of exhausting experiments to see how the theory works in practice, which might take years, as English engineer Frank Whittle found developing his jet engine. Or it appears, as the pioneering English economist John Maynard Keynes described, like a "gray woolly monster in my head."

When you look at a range of different inventions and inventors, as you can do side by side in this book, it becomes clear that, to be successful, inventions also have to come along at the right moment. Of course, as with any rule, there are exceptions to this. The mechanical clock, invented by the Chinese in the eighth century, failed to spread to the rest of the world until the twelfth century. The use of paper money, pioneered by Mongol ruler Kublai Khan in the twelfth century, did not reach Europe until the end of the seventeenth century. But generally speaking, successful inventions hit the mark because they are developed at a time when there is both the knowledge to make them possible and a need they can meet. This was true of Watt's steam engine, and also of the steam boat, plastic, and the internal combustion engine.

This raises an important question—were the inventors of these things the only people who could have made the breakthrough they did? Or did they happen to be in the right place at the right time, so they could steal a few years or days on their competitors? If Watt had died young or become caught up in the aftermath of the efforts of Charles Edward Stuart ("Bonnie Prince Charlie") to regain the throne in 1745, and never ended up as a student in Glasgow, nor met his friend the physicist John Robison, or the other doyens of the Scottish Enlightenment, would somebody else have made the breakthrough he did with steam power?

It is impossible to be sure. But let's keep a place for the role of individual genius. The inventors in this book are all people who managed to understand their field so completely that they were able to see what else, which new development, was possible. Yet, at the same time, they brought some insight from outside the field to bear on the problem. Thomas Edison, perhaps the country's greatest inventor, famously claimed that invention was 1 percent inspiration and 99 percent perspiration, and he went on to prove it in his tireless search for the right material to maximize the brightness and longevity of the filament in his lightbulb. He had more than 1,000 patents to his name, so he may have proved this point—but there are other examples to the contrary. Nor do many people really believe that they might become an Edison if they just put in enough perspiration. No, moments of inspiration are vital, too. Progress is underpinned, whether it is in technology, economics, or social development, by moments of personal imagination.

In addition to all this, inventions need wider networks of people to make the breakthrough idea a reality. They need connections with people powerful or wealthy enough to make change happen. This may explain why so many inventors are white men, because those critical networks may not have been available to anyone else. This book includes two outstanding twentieth-century women inventors. There must be many more whose graves are not marked and whose memories are not celebrated.

From the mind of a genius
*Some great inventions,
such as the X-ray, came
about by chance, but their
inventors needed the brains
and insight to realize their
life-changing significance.*

How This Book Works

There are seven sections, designed to include as much of the science, technology, and economics of inventions as possible. The first section looks at **Materials**, from cement to petroleum and plastic. **Construction & Engineering** investigates key mechanical breakthroughs such as the mechanized clock, the internal combustion engine, and the plow. In **Transport & Location**, we cover inventions connected to movement, from the pneumatic tire to GPS satellite positioning. **Medicine & Health** probes the medical field, from antibiotics to pacemakers. **Communications** runs the gamut from the printing press to the smartphone—including, of course, Tim Berners-Lee and the Internet. **Economics & Energy** looks at everything from credit cards to solar energy. Finally, **Daily Life** highlights those inventions that many people now take for granted, from heating and lighting our homes to preserving food by canning and refrigeration.

Each entry in this book is accompanied by a pithy **3-second survey**, which boils the facts down to the absolute basics. There is also a **3-minute overview**, which tries to give a different and more challenging edge to the story, together with biographical facts for the key players.

This is not a book that has to be read from cover to cover. You can dip in here and there to check facts, follow the narrative of an individual's struggle against the odds, or even derive inspiration from a great inventor's flash of insight that changed the world.

MATERIALS

alloy Mixture of two elements, which sometimes has characteristics shared by neither of its component elements.

ancient Greeks Civilization of competing and cooperating city-states in Greece, dedicated to learning and sporting prowess, which thrived for 1,000 years, c. 500 BCE–c. 500 CE.

Assyrian empire Sprawling ancient empire in the Middle East, based around the cities of Assur and Nineveh, which thrived 2300–600 BCE.

Babylonian empire Ancient empire based around the city of Babylon in what is now Iraq, which rose and fell c. 1800–c. 500 BCE.

Bakelite The first successfully and widely marketed form of plastic.

bitumen Thick, oily mixture used for fixing roofs and making roads.

concrete Mixture of cement, stones, and other aggregates that hardens over time.

Crystal Palace Glass structure designed by greenhouse designer Joseph Paxton for the Great Exhibition in London, England, in 1851 in what is now Kensington Gardens. The structure was moved south of the river, where it burned down in 1936, but not before it had inspired a new generation of modern architects.

Eddystone Lighthouse There have been four lighthouses on this spot more than 14 kilometers (9 statute miles) south of Rame Head, Cornwall, southwest England, designed to warn shipping away from the Eddystone Rocks. The third one, built by John Smeaton in 1759, used a new concrete that could set underwater and pioneered the design of lighthouses everywhere.

electroplating Process that uses an electrical current to cover an object in a thin coating of metal.

eunuch Man who has been castrated, which was common for officials in the Chinese court for some centuries.

French Revolution Uprising against distant, aristocratic rule that exploded in Paris, France, in 1789 and led directly to "the Terror," when the revolution turned into bloody reprisals against the former ruling elite.

greenhouse effect Phenomenon whereby burning fossil fuels leads to a kind of hothouse in Earth's atmosphere that traps solar heat, raising the temperature, melting polar ice caps, and disturbing the weather. It was first identified in 1824.

Hittites People living in what is now Turkey whose empire emerged in c. 1600 BCE.

Industrial Revolution Emergence of an industrial economy, fueled largely by available coal and steam power, which relied primarily on Scottish know-how and English finance. It had no clear beginning and end but arguably started around 1770, after Watt's steam engine made it possible, and settled down to more normal speeds of innovation around 1820.

iron ore Rocks in which iron deposits can be found and iron extracted.

natron Basis for an early form of soap, a mixture between soda ash (sodium carbonate decahydrate) and sodium bicarbonate.

nitrocellulose Highly inflammable compound originally known as gun cotton.

Pantheon Round Roman temple, completed by Emperor Hadrian in 126 CE, which still stands in Rome, Italy, together with its innovative concrete dome. It has been a Christian church since the seventh century.

papyrus Thick paper made from the leaves of the papyrus plant and used first by the ancient Egyptians.

paraffin Refined fuel or lubricant derived from oil. Also known as kerosene.

pig iron Impure brittle iron from the bottom of a blast furnace, so-called because the shape of the ingots resemble piglets.

polymer Molecule made from joining a number of small molecules together. It can be man-made or natural—plastic and rubber are both polymers.

radial tire Tire made from strips of rubber-coated plastic arranged at right angles to the direction of travel.

Sheffield steel Steel made in Sheffield, in northern England. The city came to specialize in its manufacture, largely because of its track record of making cutlery dating back to the thirteenth century.

vellum Parchment originally made from the skin of a calf.

CEMENT

The story of cement stretches

back into prehistory, when the ancient Greeks, Babylonians and, before them, the Assyrians began to experiment with binders capable of hardening to hold other materials together in a building. The Assyrians used bitumen and the Egyptians a mixture of sand and gypsum to create mortar, which was often mixed with small stones. But it wasn't until the Romans came along that the idea was perfected into a useful form of concrete that could be shaped in ways their forefathers had never dreamed possible. Roman architect Marcus Vitruvius Pollio—who died in 15 CE—set out the basic ideas in his *Ten Books of Architecture* and wrote down the ingredients for *pozzolana*, a crushed volcanic ash that allowed for cement to set underwater. Vitruvius' principles were put into practice spectacularly in the building of the Pantheon, beginning around 118 CE for the Roman emperor Hadrian, which included the largest unsupported dome in the ancient world and is still standing to this day. The vast system of Roman aqueducts used cement and concrete in ways that would have been impossible for previous civilizations. Perhaps surprisingly, one of the casualties of the fall of Rome in 476—and forgotten until the Renaissance—was the knowledge of how to make effective concrete.

3-SECOND SURVEY
Cement and concrete—a mixture of cement, sand or gravel, and water—are the foundations, so to speak, of civilization, because they are the critical binders that hold buildings together.

3-MINUTE OVERVIEW
The secrets of cement were not rediscovered until Renaissance scholars began to unearth Classical texts. But, modern breakthroughs happened, thanks to, among others, engineer John Smeaton. Smeaton researched ways to develop a cement that would set firmly within 12 hours, and between high tides, so that he could build the third Eddystone Lighthouse in the English Channel. It was successfully completed in 1759.

RELATED TOPIC
See also
GLASS
page 16

3-SECOND BIOGRAPHIES
MARCUS VITRUVIUS POLLIO
c. 80–70 BCE–c. 15 CE
Roman architect and engineer whose work *De Architectura* contains the earliest descriptions of hydraulic setting cement developed by the Greeks and Romans to build their harbors

JOHN SMEATON
1724–92
English civil engineer who developed modern cements by experimentation with strong, quick-drying mixes that could set underwater

JOSEPH ASPDIN
1778–1855
English cement manufacturer who patented the first version of what became known as Portland cement, due to its resemblance to Portland stone—the Victorian paneling of choice

EXPERT
David Boyle

Cement may seem modern but it was in use in prehistoric times.

GLASS

3-SECOND SURVEY

So simple a material that it can be made from sand alone, glass keeps out weather, can be used for drinking vessels and decorative objects, in lenses, and for much more.

3-MINUTE OVERVIEW

Roman manufacturers developed clear window glass (although their early versions were tinted green) and were able to mass-produce glass cups and bottles. But as with so many of their technological innovations, Roman techniques died out in the Dark Ages and had to be rediscovered. It was Italian and Bohemian merchants who created and developed the glass that dominated life in the fourteenth century and afterward.

The volcano known as Vesuvius

famously erupted in 79 CE, covering the cities of Pompeii and Herculaneum in southern Italy in volcanic ash. When the site was rediscovered, archaeologists found that some of the wealthier households in both cities had glass windows, which kept the weather out and still let in light, although unfortunately were not able to keep out the burning ash. It was a sign that, although the Romans were not the first to develop glass, they certainly drove forward its manufacture. Glass had been developed in the civilizations of the Bronze Age, including those in Syria, China, and India, but especially ancient Egypt. The pharaohs oversaw the development of glass and, well into the Roman period, the ancient city of Alexandria produced the best-quality glass in the world. When the Emperor Hadrian visited there in the second century CE, he was so impressed that he sent a translucent glass vase back to his friend Servianus in Rome. And if it wasn't the Egyptians, it was the Phoenicians, who, according to the Roman writer Pliny the Elder, claimed that they discovered the phenomenon when their merchants cooked on sand under Mount Carmel, with their pots set on top of their cargo of *natron* (a natural preservative), and found, when they finished heating their food, that glass had formed underneath.

RELATED TOPICS

See also
CEMENT
page 16

THE OPTICAL LENS
page 84

3-SECOND BIOGRAPHIES

GEORGE RAVENSCROFT
1632–83
English glass manufacturer and exporter who first discovered that adding lead to glass made it clearer and easier to shape

ROBERT LUCAS CHANCE
1782–1865
English glass manufacturer who, aided by the French glass technician Georges Bontemps, perfected the techniques for making plate glass

JOSEPH PAXTON
1803–65
English gardener whose design for London's Crystal Palace in 1851 shaped the way we use glass in buildings today

EXPERT

David Boyle

Glass illuminates our daily life, and enables scientists to see what lies beyond our world.

PAPER

RELATED TOPIC
See also
THE PRINTING PRESS
page 96

Cai Lun, an official and a senior eunuch in the court of the Chinese emperor He of Han and his successors, is credited with inventing paper and the papermaking process in around 105 CE. He was reputedly inspired by watching paper wasps make their nests out of dried vegetable fibers. Cai Lun was involved in bitter internal disputes and jockeying for power as he backed the interests of the empress Dou and her entourage. At court, he rose to be in charge of manufacturing weapons and, in that role, is credited with first creating paper, which he managed using tree bark, old rags, and fishing nets. He subsequently drank poison when the new emperor ordered him to report to prison for backing the wrong side. But Cai Lun was revered in the next century when papermaking spread rapidly through China, accelerating the growth of the literary culture that helped the Chinese to develop politically and culturally. It was not for another 500 years, when papermakers were captured by Arabs in the seventh century, that papermaking techniques were taken back to the Arab world, and then via returning crusaders, eventually reached western Europe.

3-SECOND SURVEY
Before paper, anything written down had to be scrawled onto expensive papyrus or vellum. The invention of cheap paper meant that ideas could be disseminated more easily.

3-MINUTE OVERVIEW
By 751 CE, the Chinese Tang dynasty had spread as far as the edge of the Gobi Desert. That year the Battle of Talas River halted their expansion and handed control of the Silk Road to the Muslim world. It was then that captured Chinese papermakers were said to have taken their techniques to the Mediterranean, although it took another 500 years for them to reach western Europe.

3-SECOND BIOGRAPHY
CAI LUN
48–121 CE
Senior Chinese eunuch and professional inventor who is credited with inventing modern papermaking

EXPERT
David Boyle

The nests of the paper wasp, constructed from plant material, provided inspiration for the development of papermaking techniques in China in the first century.

July 31, 1923
Born in New Kensington, near Pittsburgh, Pennsylvania

1933
Kwolek's father dies. She credited him with introducing her to the wonders of the natural world

1942
Begins her university studies at what is now Carnegie Mellon University, choosing chemistry because she can't afford the fees for medicine

1946
Offered a job with DuPont

April 1959
Her award-winning paper describing "the Indian rope trick" is published, explaining a classroom experiment to make nylon in a jar

1964
Carries out the experiment that led to the discovery of Kevlar

1971
Kevlar is marketed by DuPont for the first time. It has since earned billions of dollars for the company

1986
Retires from Du Pont having led the research team on polymers for some years

1995
Awarded the DuPont company's Lavoisier Medal for outstanding technical achievement

June 2014
The millionth bulletproof vest using Kevlar is produced

June 18, 2014
Dies at the age of 90

STEPHANIE KWOLEK

It was one of those moments of serendipity, and happened when industrial chemist Stephanie Kwolek was working as part of a team at chemical giant DuPont's research laboratory in Wilmington, Delaware. It meant that she stumbled on a new kind of plastic fiber five times stronger than steel. The result was marketed as Kelvar, and was used all over the world—best known as the lightweight basis for bulletproof vests.

"I knew that I had made a discovery," Kwolek said later. "I didn't shout 'Eureka,' but I was very excited, as was the whole laboratory excited, and management was excited, because we were looking for something new, something different, and this was it."

She was supposed to have been looking for a light fiber that could replace the steel in radial tires. That involved changing a series of carbon-base molecules into larger ones known as polymers. In 1964, she was trying to convert one of these into liquid form and it came out disappointingly muddy. But she carried on, persuaded her colleagues to risk it in the spinning machine and found that the liquid produced an extremely stiff fiber. Not only did it turn out to be far stronger than steel of equal weight, but it also resisted fire.

Kelvar began as "Fiber B," a new ingredient for tires. Aside from protective clothing, the fiber is now found in a whole range of products, including aeroplanes, mobile phones and sails.

The daughter of working-class Polish immigrants, Kwolek wanted to be a doctor but couldn't afford the tuition for medical school. She studied chemistry at what is now Carnegie Mellon University and looked for a temporary job in chemical research to fund her medical studies. She was interviewed by her future mentor Hale Charch—one of the pioneers who had just produced nylon—and, hearing that she already had a job offer elsewhere, he offered her a contract on the spot.

It was not until 1975 that Kevlar itself hit the market, and Kwolek's invention earned billions for DuPont. She made nothing out of it directly, because she had signed over her patents to the company.

David Boyle

IRON

About 5 percent of Earth's crust is composed of iron ore—the rocks and minerals that contain iron in its natural state. "Smelting" is the process of extracting the pure iron from its ore. Who first smelted iron? No one knows exactly, but the smelting process seems to have been invented toward the middle of the Bronze Age by the Hittites, a civilization in modern Turkey that reached its height in the fourteenth century BCE. When the Hittite empire fell, in c. 1180 BCE, their methods of smelting iron began to spread more widely to southern Africa and India, and then Greece and China. But the key to modern iron was the development of cheap production methods in the Industrial Revolution, including those for producing steel, which made possible so much of the rest of industrialization. One man in particular made this a reality: an English Quaker farmer's son, Abraham Darby, who in 1709 invented a blast furnace that used coke instead of charcoal in order to reach temperatures suitable for producing cast iron in major quantities. By the time Abraham Darby's grandson took over the business, they had made enough to construct the very first large-scale human structure to be made entirely of iron—the iron bridge across the Severn River in Shropshire, England—which gave its name to the town where it was built.

3-SECOND SURVEY
Iron is one of the most common minerals on Earth. Techniques to extract it, purify it and use it on a large scale made possible the Industrial Revolution in the eighteenth century.

3-MINUTE OVERVIEW
During the reign of the fourteenth-century English monarch Edward III, iron was considered so valuable that a number of the iron kitchen pots in his household were categorized as jewelry. The medieval period produced new furnaces that allowed for new cast iron shapes to be produced, with the royal pots being prime examples.

RELATED TOPIC
See also
STEEL
page 24

3-SECOND BIOGRAPHIES
ABRAHAM DARBY
1678–1717
English Quaker pioneer of modern iron smelting

THOMAS FARNOLLS PRITCHARD
1723–77
English architect and interior designer, who designed the world's first all-iron bridge at what is now Ironbridge, a town in the UK

JAMES BEAUMONT NEILSON
1792–1865
Scottish inventor who developed the hot blasting process that increased the efficiency of iron smelting

EXPERT
David Boyle

The process of smelting iron was mastered during the Bronze Age, but it required steam power to make iron smelting possible on an industrial scale.

STEEL

The Tay Bridge Disaster of 1879 occurred when a wrought iron bridge over the Tay River in Scotland collapsed during a storm, plunging a packed passenger train into the water below. It was the last in a series of similar disasters involving wrought iron bridge collapses. Engineers knew that wrought iron was not the most suitable material for building bridges and other large-scale structures. They needed something purer, harder, and stronger. But steel, which had been made as an iron alloy since at least 1800 BCE, led by the Tamils in India, was too expensive to make in large quantities. The man to change that was Henry Bessemer, whose father had worked on iron alloys at the Paris mint before fleeing to England during the French Revolution of 1789. Bessemer developed a method that involved blowing oxygen through molten pig iron to burn off impurities, which he unveiled in Cheltenham in 1856. In the audience was James Nasmyth, a Scottish engineer who had been working on the same problem; he abandoned his solution once he heard Bessemer was thinking along the same lines. Bessemer offered him a third of the patent but was turned down because Nasmyth wanted to retire. The city of Sheffield's link with steel was established when Bessemer launched a steel foundry there to produce steel using his new method.

3-SECOND SURVEY
An iron-carbon alloy, steel may not be the strongest substance known to humankind, but it is the toughest and hardest practical material that can be produced cost-effectively.

3-MINUTE OVERVIEW
In 326 BCE Porus, king of a region of India that is now the Punjab, presented Alexander the Great with a steel sword. The Tamils in Sri Lanka and southern India seem to have led the way with steelmaking in the ancient world—even though the oldest archeological evidence of steel production is actually in Anatolia (modern day Turkey) from c. 1800 BCE.

RELATED TOPIC
See also
IRON
page 22

3-SECOND BIOGRAPHIES
HENRY BESSEMER
1813–98
English engineer, originator of the Bessemer process and many other inventions, including a ship that kept passengers level so they never experienced seasickness

ANDREW CARNEGIE
1835–1919
Scots-American who made a fortune from steel production and became, at one time, the richest man in the world

EXPERT
David Boyle

High tensile strength and low manufacturing costs enable steel to be widely used in major construction projects, from skyscrapers to suspension bridges, railroads to rivets.

PETROLEUM

In 1847, Scottish chemist James Young noticed that black oil was seeping out of the roof of a colliery in Derbyshire, England. He began experimenting with the oil and managed to produce a lighter fuel oil for lamps and a heavier oil for lubricating machinery. Four years later, the mine's supply of oil had begun to dry up, but Young discovered it had been seeping through the sandstone roof of the mine and realized that it might be possible to create it artificially. It was this idea that led him to produce what he called "paraffine oil," or paraffin. The invention made his fortune, eventually leading to the mining of shale oil deposits in Scotland to produce it. Before his great breakthrough, Young had educated himself at night school in Glasgow and became close friends with the Scottish missionary David Livingstone. After his discovery, Young spent his money on yachting, scientific experiments, and in funding Livingstone's African explorations. The word "petroleum" derives from the Latin word for rock, and "rock oil" was the phrase used to describe the substance that comes out of the ground, then mainly from Baku in Russia, where the first oil refinery was built in 1861.

RELATED TOPIC
See also
INTERNAL COMBUSTION ENGINE
page 48

3-SECOND SURVEY
Petroleum products now dominate the world economy, not always to the advantage of the nations that supply the oil. Economists call the side effects of these bonanzas the "Curse of Oil."

3-MINUTE OVERVIEW
The realization that oil could be refined into fuel and other necessities turbocharged twentieth-century economics, but it has also contributed to raising the overall temperature of the planet, changing the climate, and boosting the so-called "greenhouse effect'—a phrase coined in 1917 by Scottish inventor Alexander Graham Bell.

3-SECOND BIOGRAPHIES
JAMES YOUNG
1811–83
Scottish chemist, inventor, and philanthropist who first made paraffin and is credited with the breakthroughs that allowed petroleum to be refined

JAMES MILLER WILLIAMS
1818–90
Canadian businessman, railroad manufacturer, and politician who launched the petroleum industry in Canada, and pioneered similar businesses across North America

JAN JÓZEF IGNACY ŁUKASIEWICZ
1822–82
Polish pharmacist who built the first working oil refinery and invented the kerosene lamp

EXPERT
David Boyle

Mining oil transforms landscapes—just as it has transformed economic fortunes.

PLASTIC

Metallurgist Alexander Parkes

was an expert in electroplating before he was credited with inventing the first man-made plastic material, which he called Parkesine. Patented in 1862, his invention treated cellulose with nitric acid and a solvent to create what was actually nitrocellulose—now popularly known as synthetic ivory. He went into business making his material in a factory in Hackney, in east London, but the venture failed, because Parkesine was highly flammable, cracked easily, and was relatively expensive to produce. Some of Parkes's 17 children helped in his experiments, which tried to replicate some of the properties of natural materials such as rubber or ivory. But Parkesine was less successful than its later rivals, xylonite and celluloid, which took its place in the latter years of the nineteenth century. All three were then overtaken by the arrival of Bakelite in 1907. The invention of Belgian-American chemist Leo Baekeland, Bakelite was the first completely synthetic plastic material and appeared just in time for it to be used widely during World War I. The full chemical name of Bakelite was polyoxybenzylmethylenglycolanhydride, and it had magical insulating properties that meant it was unable to conduct heat or electricity. This made Bakelite perfect as a material for telephones and other electrical gadgets.

RELATED TOPIC
See also
THE CREDIT CARD
page 130

3-SECOND SURVEY
Before the invention of plastics nearly all household objects were made from natural products such as bone or wood; plastics are much harder wearing and can be mass-produced cheaply.

3-MINUTE OVERVIEW
Bakelite inventor Leo Baekeland faced legal action to defend his patents in the United States, and winning his case led, in 1922, to a merger between three companies—his own included—to form the Bakelite Corporation. The trademark they used incorporated the mathematical symbol for infinity, and they sold their product aggressively under the slogan: "The material with a thousand uses."

3-SECOND BIOGRAPHIES
ALEXANDER PARKES
1813–90
English metallurgist who invented Parkesine

LEO BAEKELAND
1863–1944
Belgian-American chemist who began experiments after selling his photographic paper patents to Eastman Kodak and promising not to work on photographic products for 20 years. The result of his experimentation was Bakelite

EXPERT
David Boyle

The Bakelite telephone was one of the first plastics found in people's homes.

CONSTRUCTION & ENGINEERING

CONSTRUCTION & ENGINEERING
GLOSSARY

astrolabe Astronomical instrument designed to measure the position of stars and planets in the sky, as an aid to navigation.

bilges Lowest compartment of a ship.

carbon dating Test designed during the 1940s to work out the age of an item by measuring the amount of carbon-14 (a radioactive isotope of carbon) it contains. By using the half-life of the isotope, and measuring how much it has decayed, it is possible to calculate the age of the object.

gas turbine Combustion engine that can turn fuel into power.

Hanging Gardens of Babylon One of the Seven Wonders of the Ancient World, and reputed to include gardens hundreds of feet high, in the ancient city of Babylon and built by Nebuchadnezzar II in c. 600 BCE.

ironmonger Purveyor of tools, nails, and other utensils. The British term for a hardware dealer.

London Blitz Sustained bombing campaign against London carried out by the German Luftwaffe that lasted from September 1940 to May 1941. Although destructive, it did not cause as much devastation as the later bombing of German and Japanese cities by British and American air forces.

Mesopotamia Cradle of civilization in the Bronze Age, the area between and around the Tigris and Euphrates rivers in what is now Iraq and surrounding areas.

propeller Type of fan that pushes itself through the air by turning its rotation into a forward or backward thrust.

Scottish Enlightenment Hugely creative period in Scottish history that followed the 1745 uprising (an attempt to regain the throne for the House of Stuart) and led to a range of important inventions. Scottish Enlightenment continued through technological know-how well into the twentieth century and the invention of the television by the Scotsman John Logie Baird (see page 106).

seed drill Piece of agricultural equipment that inserts seed into the ground at the correct depth and covers them with soil, so that they are not consumed by birds.

steam Gaseous form of water, which has immense power to drive machinery.

Syracuse Greek city-state founded on the island of Sicily and immensely powerful in the ancient world.

Tang dynasty Period in China from 618 to 907 CE, usually thought of as the high point of Chinese civilization.

Ur Great Sumerian city in Mesopotamia, once at the mouth of the Euphrates River (although the coastline has since moved) and which rose to prominence in c. 3600 BCE.

vacuum Absence of air or other gases.

THE WHEEL

Who was it who first invented the wheel? This is important because it is, in some ways, the most significant discovery ever made by human beings. The trouble is that the answer is lost in the mists of time. You might as well ask which human being first stood upright on two legs. Not even the experts agree, but there is a prevailing opinion that the first wheels were actually potter's wheels, in Mesopotamia, sometime around 3500 BCE. The earliest example to be found by archaeologists was unearthed at the ancient city of Ur, in what is now Iraq. Carbon dating suggests that it was made around 3129 BCE. But it is confusing that, in fact, wheels appeared at the same period all over the world, from China and India to Europe and ancient Greece—the ancient Greeks also invented the wheelbarrow. The first image of a wagon with obvious wheels, dated to 3330 BCE, was dug up in Poland. What seems to have turned potter's wheels into transport and locomotion was the domestication of horses. Some bright spark in one of these places saw the potters at work making drinking vessels, and made the connection with the horses outside. The rest is history, but we will never know who was responsible.

3-SECOND SURVEY
Wheels sprung up simultaneously in many parts of the world, enabling people to move heavy objects vast distances.

3-MINUTE OVERVIEW
Although several cultures could claim responsibility for inventing the wheel, there are some that clearly can't. The ancient American civilizations did come up with a wheel, but they used it as a children's toy. The difference was that, unlike in Asia and Europe, the original Americans had no horses, and buffalo did not take kindly to domestication and pulling carts.

RELATED TOPICS
See also
THE STEAM ENGINE
page 46

THE INTERNAL COMBUSTION ENGINE
page 48

3-SECOND BIOGRAPHY
LEONARD WOOLLEY
1880–1960
English archaeologist who led excavations at Ur in Mesopotamia, where the oldest surviving potter's wheel was found

EXPERT
David Boyle

The invention of the wheel as a means of transport marks a turning point in the development of human evolution.

THE NAIL

Nails were luxury items well into

the nineteenth century. In the fledgling United States, unable to access the nails that were made in such quantities in England, people used to make them during the long winter evenings around the fire. President Thomas Jefferson used to make nails when he had nothing better to do. Often, when people moved, they first used to remove the nails because they were so important for holding together the joists of their new home. What made the difference was the invention of "cut nails," in square shapes made from wrought iron by machine, a process pioneered by the Massachusetts goldsmith Jacob Perkins. Perkins was extraordinarily inventive, pioneering refrigeration and a forerunner of the machine gun, known as the "steam gun" and rejected by the Duke of Wellington on the grounds that it was "too destructive." Perkins had a rival in England called Joseph Dyer, the son of a captain in the Rhode Island Navy, who set up in Birmingham in England with his own nail-making machine and eventually lost a fortune in the collapse of the Bank of Manchester. Both innovations were overtaken by the development of so-called "wire nails," which provided us with the cheap, plentiful, and diverse nails that we have today.

3-SECOND SURVEY
Nails make it possible to put two pieces of wood together in a way that will hold them in place for a lifetime—or longer.

3-MINUTE OVERVIEW
Who actually invented nails? Nobody knows, but as with so many other pioneering discoveries in this book, they seem to have emerged in the Bronze Age, and to have come from the Middle East. The earliest bronze nails found have been carbon-dated back to around 3400 BCE and come from ancient Egypt. They also seem to have been relatively expensive.

RELATED TOPICS
See also
IRON
page 22

STEEL
page 24

3-SECOND BIOGRAPHIES
JACOB PERKINS
1766–1849
Pioneering American goldsmith who invented the process of manufacturing cut nails

JOSEPH DYER
1780–1871
American-born inventor, and a manufacturer of cut nails

JOSEPH HENRY NETTLEFOLD
1827–81
British industrialist responsible for bringing wire nails onto the market in vast quantities, and manager of a company now known as GKN

EXPERT
David Boyle

Over the millennia, nails have been made from bronze, brass, iron, copper, and aluminum, but today the most commercially viable option is steel.

ARCHIMEDES' SCREW

The Archimedes screw—one of the oldest machines still in use—is a device designed to lift low-lying water for drainage or irrigation. More accurately described as a screw pump, the invention is made up of a screw-shape cylindrical shaft, made of metal or wood, inside a hollow pipe. The screw was turned either by hand or by windmills—the invention was widely used in the Netherlands to drain and reclaim land. As the shaft is turned, the rotating screw shape scoops up water and pushes it out the top of the tube. The Greek engineer and polymath Archimedes is reputed to have come up with the invention to save a ship he had designed. The *Syracusia* was an enormous boat, part luxury vessel, part warship, and big enough to hold a garden, gymnasium, and temple to the goddess Aphrodite—but its hold was filling with sea water. Archimedes' simple mechanical pump saved the day, although his ship sailed only once, from Syracuse to Alexandria. Archimedes' design proved to be more long lasting. It is still found in uses as diverse as sewage treatment plants and chocolate fountains and it stabilizes the leaning Tower of Pisa in Italy.

3-SECOND SURVEY
Archimedes managed to design a simple pump for the ancient world that is still in use today, particularly for sewage.

3-MINUTE OVERVIEW
Did Archimedes actually think up the screw himself, or did he just see one in action in Alexandria and bring it home? There were reports that the water was raised in this way to irrigate the Hanging Gardens of Babylon, and the classicist Stephanie Dalley has found an Assyrian inscription that seems to push the origin of the screw back to 350 years earlier.

RELATED TOPIC
See also
THE WHEEL
page 34

3-SECOND BIOGRAPHY
ARCHIMEDES
287–212 BCE
Pioneering engineer and mathematician, famous throughout the ancient world, born in Syracuse, Sicily, and murdered there by a Roman soldier

EXPERT
David Boyle

The Greek engineer Archimedes was killed for staring at a set of circles he had drawn on the ground instead of obeying a Roman soldier's commands, but his simple pump design lives on.

June 1, 1907
Born in Coventry, England

August 27, 1928
Joins 111 Squadron and
wins a reputation for his
aerobatic displays

January 27, 1936
Enters agreement to set
up Power Jets Ltd,
formed to develop the
first jet

August 27, 1939
The first jet-power plane,
the Heinkel He 178,
makes its first flight in
Rostock

December 10, 1940
US patent is applied for;
the strain becomes too
much for Whittle, who
has a breakdown and
is absent from work for
a month

May 15, 1941
The first British jet flies,
at RAF Cranwell in
Lincolnshire; the first
flight of the American
XP-59A Airacomet
follows in October 1942

August 26, 1948
Retires from the RAF on
medical grounds. He later
lives in the United States

August 9, 1996
Dies in Columbia,
Maryland

June 1, 2007
A statue of Whittle is
unveiled in his birth town
of Coventry

FRANK WHITTLE

The British are famous for being innovative but they sometimes find it difficult to break through an innate conservatism—an adherence to the accepted way of doing things. This means that the story of British invention is often one of brilliant ideas getting stuck. And this was certainly the story of the invention of the turbojet engine by aircraft engineer Frank Whittle.

Born in 1907 in Coventry, England, Whittle was the son of a mechanic. He was determined to join the Royal Air Force, but because he measured only 5-feet tall, he kept on being turned down despite passing the entrance exams with flying colors. He was eventually accepted on his third attempt in 1923.

From his time as an air cadet onward, Whittle was fascinated by the problems of speed and how to make existing engines go much faster without a necessary increase in the weight and complexity of the propeller engines that were used then. Planes would also need to fly higher to reduce the air pressure, and the existing engine design wasn't suitable for that, either. He came to the conclusion that a gas turbine might provide the necessary power for a major breakthrough in speed.

The result was his design for a turbojet, a fan that sucked in air, squeezed it, and set it alight, then pushed it outward with great power. After the air ministry said it was "impractical" he patented the idea himself and, because the RAF didn't want it, it was never a state secret. By 1935, he had managed to attract financial backing and started a company called Power Jets Ltd. This was slow going and the official approvals or finance he needed was even slower, while in Germany the Nazis were also pressing ahead with a similar idea. In fact, Whittle was told later that if the German leader Adolf Hitler had known there was a man in England designing planes that could travel at 500 mph (800 km/h), World War II might never have happened.

When Whittle's first turbojet flew in May 1941, immediately after the end of the London Blitz, he had been beaten by 21 months by German pioneer Ernst Heinkel. Five months later, the Bell Aircraft Corporation in America started development of the Airacomet; it was airborne in October 1942, whereas the British Meteor jet did not take to the skies until 1943.

David Boyle

THE PLOW

RELATED TOPICS
See also
IRON
page 24

THE STEAM ENGINE
page 48

3-SECOND SURVEY
A plow breaks up the soil in a field so that the seeds can be sown; it is one of humankind's first inventions.

3-MINUTE OVERVIEW
The first plow emerged almost as soon as people began to stay put long enough to till the soil. They were sticks pulled by animals, or before that, just handheld tools to open up the soil. From the ancient Egyptians to the inventor of the light tractor, Harry Ferguson (1884–1960), in the 1920s, the story of the plow has been a story of small steps forward.

Probably the man who did most to revolutionize what is, after all, one of the most ancient human inventions of all was a young technician from Berwickshire in southern Scotland called James Small. He was working in Yorkshire, England, when he was 18 and caught sight of one of the iron "Rotherham plows" invented by Joseph Foljambe, who worked in the area and had pioneered the first commercial plow made of metal in 1730. Even so, it took about three decades for his invention to percolate out of the Rotherham region of Yorkshire, where Small saw it. Small took the idea home to Scotland and, in a series of experiments to find the best shape and most appropriate metal, he created what was known as the "Scots Plow," which was soon exported to the United States. He also refused to patent the idea, because he thought his invention should be freely available, although this led at one stage to him spending time in a debtors' prison. Between them, the efforts of Foljambe and Small meant that the business of plowing a field could be done by a single plowman and his horse, instead of several men and a team of oxen.

3-SECOND BIOGRAPHIES
JETHRO TULL
1674–1741
English farmer who invented the seed drill, allowing seed to be evenly spaced, and launched the "new husbandry" on the world—much to the horror of his laborers

JOSEPH FOLJAMBE
fl. 1730
English inventor of the first iron plow, or Rotherham plow, the forerunner of the other improvements

JAMES SMALL
1740–93
Scottish perfecter of the metal plow, who refused to patent his discovery

EXPERT
David Boyle

Early plows relied on human power before the use of draft animals, then machines, to till the soil for crops.

THE MECHANIZED CLOCK

RELATED TOPIC
See also
THE COMPASS
page 54

3-SECOND SURVEY
The mechanized clock revolutionized time-keeping and enabled societies to communicate and cooperate with far greater precision.

3-MINUTE OVERVIEW
The mechanical clock, attached to tolling bells, was not an instrument of liberation in Europe. It was used to impose timetables on ordinary people—when they must get up, when they could break from work in the fields, when they had to go to bed. Timekeeping—and being punctual—became a religious duty, the onset of a long alliance between puritans and clocks.

Yi Xing, one of the most brilliant inventors in history, was a mathematician, an engineer, and a Buddhist monk during the Chinese Tang dynasty. He was also an expert in astronomy and the calendar; he built a series of 13 observatory posts, stretching from Vietnam to Siberia, to perfect his calculations for solar eclipses. Working with a military engineer and government official, Liang Lingzan, his greatest invention was in 725 CE: a mechanical clock that involved water steadily dripping on a wheel that made a full revolution every 24 hours. It combined water power with cogs and wheels and was built in the shape of a great metal sphere that tracked the movements of the stars and planets. The invention may have been a clock, but it did not tell the immediate time. Nor did the eleventh-century mercury-powered astrolabe, invented by a mechanical engineer, al-Murādī, although the device made its way to the Iberian peninsula, where al-Murādī lived, ready to inspire the revolution in mechanical clockmaking from around the 1260s onward. This was when every abbey and cathedral vied with each other to make the most spectacular, ingenious clock mechanisms—mainly powered by more reliable weights. At this time, however, they did not have clock faces; the hours were chimed using bells.

3-SECOND BIOGRAPHIES
YI XING
683–727
Chinese astronomer who made what seems to have been the first mechanical clock, though still powered by water

LIANG LINGZAN
fl. 720s
Chinese official who worked with Yi Xing to make the first mechanical clock

ALĪ IBN KHALAF AL-MURĀDĪ
fl. c. 1050
Mathematician from Arab Spain who made the first European mechanical clock using mercury

EXPERT
David Boyle

Compared with the sundial or hourglass as a means of telling the time, mechanical clocks enabled more accurate timekeeping in the medieval world.

THE STEAM ENGINE

James Watt was not actually the inventor of the steam engine. Like so many other engineers who emerged from the Scottish Enlightenment, he took an existing idea and made it work effectively. The original inventors were two Englishmen: a military engineer, Thomas Savery, who first invented an "engine for raising water by fire," and Thomas Newcomen, an ironmonger, whose machine was used for pumping water out of flooded mines. Watt was born in Greenock on the Clyde River in western Scotland. His grandfather was a professor of math and his father made nautical instruments for the herring fleet. He was hardly poor, but the young Watt still had his difficulties. He was famously moody and difficult to get on with. He began mending and then improving the scale model of Newcomen's engine that Glasgow University used for lectures. He found it was not producing enough steam for it to work efficiently. The piston turned only two or three strokes at a time and most of the steam escaped, and this irritated him. Determined to find ways to prevent the steam in the Newcomen engine being wasted, he made a crucial breakthrough in April 1765, when he realized that steam would rush into a vacuum and need not cool the cylinder, as was the case with Newcomen's engine.

3-SECOND SURVEY
The sheer power of steam made so much else possible. It could drive locomotives, factories, ships, agricultural machinery, and much more besides—and soon it did.

3-MINUTE OVERVIEW
The main difference between Watt's and Newcomen's engine was that the steam actually drove the piston instead of creating a vacuum. A few days after Watt's breakthrough, his friend the physicist John Robison barged into his workshop and realized that Watt had solved the problem. But he saw his nervous friend kick the experiment under the table with his foot, terrified that Robison would let the secret out.

RELATED TOPIC
See also
THE INTERNAL COMBUSTION ENGINE
page 48

3-SECOND BIOGRAPHIES
DENIS PAPIN
1647–1713
French physicist who invented a forerunner of the steam engine—the pressure cooker

THOMAS SAVERY
1650–1715
English engineer who built the world's first steam engine

THOMAS NEWCOMEN
1663–1729
English ironmonger who built a steam engine using a vacuum created by condensed steam

JAMES WATT
1736–1819
Scottish inventor whose improvements to the steam engine proved a breakthrough

EXPERT
David Boyle

The inventors of the steam engine needed to understand atmospheric pressure and the nature of a vacuum.

THE INTERNAL COMBUSTION ENGINE

3-SECOND SURVEY

The horseless carriage developed into a must-have element of modern life, which seemed to set people free to come and go as they pleased.

3-MINUTE OVERVIEW

This invention turned out to be a double-edge sword. There are now well over a billion cars on highways of the world, leading to congestion and pollution that shortens the lives of the poorest people, who tend to live closest to the traffic. Nor can the internal combustion engine set people free anymore—all too often, it condemns the owners to long, exhausting journeys on congested freeways, breathing in the fumes.

France was the home of the early pioneers of the internal combustion engine—the basic workings of which became the automobile. The invention drives an engine by managing small explosions of gasoline in a combustion chamber inside the mechanism. It was developed almost simultaneously by two Frenchmen, one on land and the other on water. The first was a retired soldier, inventor, and politician, François Isaac de Rivaz, who was at that time living in Switzerland. It was de Rivaz who first built an engine along these lines, powered by a mixture of oxygen and hydrogen and ignited by an electric spark; he patented the idea in 1807. The following year he built a "horseless carriage" for the engine to drive. At the same time, Joseph Nicéphore Niépce and his brother Claude were building a similar engine, called the Pyréolophore, fueled by a mixture of crushed coal dust, moss, and resin. This powered a boat on the Saône River and won them a patent the same year (1807). The first modern car was the 1860 brainchild of the Belgian Étienne Lenoir, living at the time in Paris—an adopted Frenchman at least—after which the story passes to Germany.

RELATED TOPIC

See also
THE STEAM ENGINE
page 46

3-SECOND BIOGRAPHIES

FRANÇOIS ISAAC DE RIVAZ
1752–1828
Retired French army officer behind the design for the first hydrogen-powered internal combustion engine

JOSEPH NICÉPHORE NIÉPCE
1765–1833
Pioneer French photographer who developed the first internal combustion engine on a boat, patented in 1807

ÉTIENNE LENOIR
1822–1900
Belgian engineer who built the first commercial automobile engine

EXPERT
David Boyle

Efforts to build more environmentally friendly vehicles may spell the end of the internal combustion engine for future transportation of goods and people.

TRANSPORT & LOCATION

alternative fuels Fuel other than traditional petroleum fuels (petrol or diesel); new technology for powering an engine includes electricity, solar or biofuels.

auto-pilot systems System used to control a vehicle, such as a car or plane, without the need for constant 'hands-on' control.

bloomers Women's underclothes with legs, popular in the nineteenth century, particularly for riding bicycles.

carbon emissions Carbon dioxide (CO_2), one of the greenhouse gases, released into the air by burning fossil fuels; driving a petrol-powered vehicle is the single largest source of carbon emissions for the typical household in high income countries.

Cold War Period of political and military tension after the Second World War between the West (the United States and NATO allies) and the Eastern bloc (the Soviet Union and its states), which lasted from 1947 until the collapse of the Soviet Union in 1991.

Communist (Russian) Revolution Two revolutions in Russia in 1917 that led to the overthrow of the tsar and the rise of the Soviet Union.

deviation Error induced in a compass by local magnetic fields caused by iron and electric currents; magnetic deviation must be allowed for to calculate accurate bearings.

Doppler effect Increase (or decrease) in the frequency of sound, light or other waves as the source and observer move towards (or away from) each other; named after the Austrian physicist Christian Doppler, who proposed it in 1842.

female emancipation Legal, political and socioeconomic rights claimed for women, equivalent to those of men, with respect to property, suffrage, employment and so on. These formed the basis of the women's rights movement in the nineteenth century and the feminist movement in the twentieth century.

fixed-wing plane Aircraft, such as an aeroplane, with wings that stay in the same position, rather than moving up and down or rotating; lift is generated instead by the craft's forward motion/airspeed and wing shape.

horsepower Unit originally defined as the amount of power that a horse could provide; now used most commonly as a non-metric unit of mechanical power.

hybrid vehicles Means of transport that use a combination of petroleum and electric power.

India rubber Another name for natural rubber, which is produced by rubber trees, mainly in India and Latin America.

Italian Renaissance Cultural rebirth of art, architecture, literature and learning that originated in Italy in the fourteenth century and later spread throughout Europe.

Korean War Conflict (1950–3) between South Korea (supported by the United States and the United Nations) and communist North Korea (supported by China and the Soviet Union). No peace treaty has ever been signed so the two countries are still technically at war.

lodestones Naturally magnetic stones that attract iron, and can therefore be used as a compass, as they were by the Chinese. There is evidence they were used a thousand years ago in central America for astrological purposes or to align temples with magnetic or magical lines of force.

magnetic north The north pole as determined by using a compass only. It moves about with changes in the magnetic fields of Earth's crust. Magnetic north is near to but distinct from geographical north.

magnetic poles The two poles, north and south, where Earth's magnetic fields point directly downwards (north) and upwards (south).

mass production Process of manufacturing large quantities of similar products efficiently, frequently utilizing assembly-line technology, as in the case of Henry Ford's automobile.

penny-farthing Early bicycle with a huge front wheel and a small back wheel. The name derives from the respective size of the penny and the farthing (worth a quarter of a penny). Its front wheels could be anything up to 150 centimetres (60 inches) in diameter, and were designed to provide speed – in a way that is now provided by gears. Penny-farthings were so dangerous that when they were replaced in the 1890s by a more recognizable design it was known as the safety bicycle.

steam Gaseous form of water, which has immense power to drive machinery.

vacuum Absence of air or other gases.

Vietnam War Conflict (1955–75) between non-communist South Vietnam (aided by the United States and later Australia, New Zealand the Philippines, South Korea, and Thailand) and communist North Vietnam, two parts of what was once the French colony of Indochina.

THE COMPASS

3-SECOND SURVEY

The compass transformed navigation because it gave reliable bearings, which made it possible to explore new oceans and lands.

3-MINUTE OVERVIEW

A magnetic compass points to the magnetic north pole, approximately 1,000 miles (1,600 kilometers) from the true geographical North Pole. By the fifteenth century, explorers had realized this discrepancy between magnetic north and true north, and made adjustments in compass calculations. Mariners also need to adjust for deviation, the response of the compass to local magnetic fields caused by iron and electric currents. Sometimes compensating magnets are placed under the compass itself.

Early navigators relied on the stars, Moon, and Sun, as well as earthbound landmarks. Problems arose when the weather was overcast or foggy. Before the compass was adopted in Europe in the late thirteenth century, sea travel was limited in the Mediterranean between October and April, because of the lack of clear skies. The compass gave a constant bearing with its magnetized pointer determining direction relative to the Earth's magnetic poles. This led to great improvements in the safety and efficiency of travel, opening up the oceans for exploration and the discovery of the New World. The Chinese are credited with the original idea, but the invention was first used for spiritual practices, not navigation. Early compasses were made from lodestone, a naturally magnetized ore of iron. Magnetized needles replaced lodestone in the eighth century, and were floated in a bowl of water. Dry compasses were developed as navigational devices on ships, appearing around 1300 in medieval Europe. The liquid-filled, handheld compass was not invented until the early twentieth century. The fluid causes the magnetized needle to stop quickly instead of move back and forth around magnetic north. Today, global positioning system (GPS) receivers have begun to replace compasses.

RELATED TOPIC

See also
GLOBAL POSITIONING SYSTEM (GPS)
page 66

3-SECOND BIOGRAPHIES

ZHENG HE
1371–1435
Chinese navigator, the first person recorded to have used the compass as a navigational aid; he made seven ocean voyages in 1405–33

TUOMAS VOHLONEN
1877–1939
Finnish surveyor who in 1935 invented and patented the first portable liquid-filled compass designed for individual use

EXPERT
Judith Hodge

Despite the invention of navigation aids using GPS, compasses remain popular because they are cheap, durable, and do not require electricity.

THE SAFETY BICYCLE

Early versions of the bicycle in the first half of the 1800s were uncomfortable wooden contraptions with steel wheels and fixed gears. Manufacturers then enlarged the front wheel to ensure a smoother, faster ride, but these "penny-farthings' were mainly the preserve of thrill-seeking young men who used them for racing. The English inventor John Kemp Starley developed the Rover "safety" bicycle in 1885. With its lower center of gravity this model greatly reduced the danger of "taking a header" over the handlebars, making braking more effective and increasing the popularity of cycling. Starley's Rover was the forerunner of the modern bicycle; its features are recognizable today, such as near equal-size wheels, a diamond-shape frame, pedals below the saddle that power the back wheel through a chain and gears, and handlebars to a forked front wheel. John Dunlop's reinvention of the pneumatic bicycle tire in 1888 contributed to a smoother ride on paved streets. At first, cycling was a relatively expensive hobby, but mass production made the bicycle a practical investment for working men getting to and from a job. Women, too, began cycling in increasing numbers, with dramatic changes in their lives. The then-shocking bloomers replaced bustles and corsets, giving a woman more mobility to ride a bike.

3-SECOND SURVEY
The safety bicycle transformed people's ability to travel, particularly working people and women, and became the most important form of personal transport.

3-MINUTE OVERVIEW
The bicycle "craze" of the 1890s sparked by the Rover's invention had an impact on female emancipation. As bicycles became safer and cheaper, more women had access to the personal freedom the machine embodied. The bicycle came to symbolize the New Woman of the late nineteenth century, especially in Great Britain and the United States. Bicycles were known as "freedom machines" by feminists and suffragists. They gave women unprecedented mobility and liberated them from restrictive garments.

RELATED TOPICS
See also
THE WHEEL
page 34

THE PNEUMATIC TIRE
page 58

3-SECOND BIOGRAPHIES
JOHN DUNLOP
1840–1921
Scottish inventor and veterinarian who reinvented the pneumatic tire for bicycles at his home in Belfast, where he had settled, in 1887

JOHN KEMP (J. K.) STARLEY
1854–1901
English inventor of the revolutionary "safety" bicycle, on which modern-day bicycle technology is still based

EXPERT
Judith Hodge

Cycling, first made popular when the safety bicycle replaced earlier models known as "bone-shakers," is today experiencing a resurgence.

THE PNEUMATIC TIRE

3-SECOND SURVEY

Invention of the pneumatic tire, providing a cushion of air between road and vehicle, played a key role in the boom in bicycles and automobiles.

3-MINUTE OVERVIEW

The name of another Scottish-born inventor, John Boyd Dunlop (1840–1921), is more closely associated with the invention and the company that still bears his name. Familiar with making rubber devices, he reinvented pneumatic tires for his son's tricycle and developed them for use in cycle racing. After a cyclist using his tires started winning all the races, Dunlop sold his rights to Dunlop Rubber, a company formed by president of the Irish Cyclists' Association, Harvey Du Cros.

The pneumatic tire was first developed by Scottish inventor Robert Thomson. He was only 23 when he patented his invention in 1846—a hollow belt of India rubber inflated with air so that the wheels presented "a cushion of air to the ground, rail, or track on which they run." The belt of rubberized canvas was encased within hard-wearing leather that was bolted to the wheel. A demonstration of Thomson's "Aerial Wheels" fitted to several horse-drawn carriages took place in Regent's Park, London, in March 1847. The pneumatic tires greatly improved both passenger comfort and noise levels, and one set was reported to last for 1,200 miles (1,950 kilometers). But Thomson's invention was ahead of its time; not only were there no automobiles, but bicycles were only just starting to appear. Lack of demand and high production costs meant that pneumatic tires remained a mere novelty for another 30 years, until their reinvention by John Boyd Dunlop. Thomson developed and patented many other inventions, including the first successful mechanical road haulage vehicle, a steam traction engine. His solid India-rubber tires (another Thomson patent) enabled heavy steam engines to travel along roads without damaging the surface. By 1870, "Thomson Steamers" were being manufactured and exported around the world.

RELATED TOPICS

See also
THE SAFETY BICYCLE
page 56

THE AUTOMOBILE
page 62

3-SECOND BIOGRAPHY

ROBERT WILLIAM THOMSON
1822–73
Scottish inventor who patented the pneumatic tire in 1846

EXPERT

Judith Hodge

Overlooked as the inventor of pneumatic tires, a vintage vehicle rally is held in honor of Robert Thomson every June at his Scottish birthplace, Stonehaven.

April 16, 1867
Wilbur born in Milville, Indiana

August 19, 1871
Orville born in Dayton, Ohio

1889–92
The brothers launch several newspapers

1892
They open bicycle repair and sales shop, selling own bicycle design

1900
They carry out first tests with glider at Kitty Hawk

December 17, 1903
Wilbur Wright makes the first controlled, sustained powered flight, in the *Wright Flyer I*

1905
The brothers develop and demonstrate the first practical airplane

1906
Patent granted on airplane control system

1908
Public demonstrations in Europe and North America. Orville crashes and is badly injured; his passenger becomes the first person to die in an airplane crash

1909
US Army buys its first military aircraft. The Wrights begin to manufacture airplanes and teach pilots; they become wealthy businessmen

May 30, 1912
Wilbur dies of typhoid at the age of 45 in Dayton

1915
Orville sells the business

January 30, 1948
Orville dies at the age of 76 in Dayton

THE WRIGHT BROTHERS

It was a childhood toy, a model helicopter, brought home by their preacher father that sparked Wilbur and Orville Wright's lifelong interest in mechanics and aviation. The brothers worked together on various enterprises, including newspaper publishing and running a bicycle shop in their hometown, Dayton, Ohio. They were swept up in the cycling craze following the invention of the safety bicycle, even designing their own model.

The Wrights followed the research of early aviators, such as Otto Lilienthal, who died in a glider crash. Experimenting with "wing warping" (a system they later patented of pulleys and cables to control a fixed-wing aircraft) based on bird flight, the brothers built and tested their flying machines at Kitty Hawk, North Carolina, known for its strong winds. It was here on December 17, 1903 that they made the first controlled, sustained flight of a powered, heavier-than-air aircraft. Wilbur flew the *Wright Flyer I* for 59 seconds, at 852 feet (260 meters), recorded in a famous "first flight" photograph.

The achievement did not bring them overnight fame and fortune. Instead, news of the event was met with great skepticism both at home and abroad: the French press dismissed them as "bluffeurs." Over the next two years, the brothers developed the first practical fixed-wing airplane, which they demonstrated in 1905. Although not the first to fly experimental aircraft, the Wrights invented the controls that made flight possible, and they patented their design in 1906.

Unable to find a market for their invention, Wilbur went to Europe where, during public demonstrations in France, he stunned onlookers with the plane's abilities and his skills as a pilot, performing figures eight. Back in the United States, similar displays by Orville resulted in a crash, leaving him badly injured and his passenger, Thomas Selfridge, dead. Shocked by the event, Wilbur pushed himself to greater lengths and set new records for altitude and duration.

The Wrights became huge celebrities, feted by royalty and heads of state, as well as wealthy businessmen with sales of their planes in Europe and the United States. Wilbur died in 1912, and Orville sold the business in 1915. The remaining Wright brother made his last flight in 1944, noting that the plane had a wingspan longer than the distance of their first flight in 1903.

Judith Hodge

THE AUTOMOBILE

Karl Benz is generally credited

with the invention of the modern automobile powered by an internal combustion engine. The Benz Patent-Motorwagen went into production in 1886, after Benz's wife Bertha made the first successful road trip from Mannheim to Pforzheim and back in the "horseless coach." Close behind came fellow German inventors Gottlieb Daimler and Wilhelm Maybach; in 1889 they designed the first vehicle from scratch as an automobile, not as a horse-drawn carriage fitted with an engine. Before this, earlier self-propelled vehicles included Nicolas-Joseph Cugnot's steam-power tractor for the French army (1769), and Robert Anderson's electric carriage (c. 1832–39). The focus of British inventors was on rail after the Locomotive Act (from 1865 to 1896, self-propelled vehicles on public roads had to be preceded by someone waving a red flag) effectively killed off auto development. It was in the United States that cars were rapidly adopted in the early twentieth century. Henry Ford's mass-production techniques led to the first affordable car for the general public, the 1908 Model T. By the 1930s, most of the mechanical technology used in cars today had been invented, although some of it was later reinvented. Future automobile inventions are focused on energy efficiency, alternative fuels, and autopilot systems.

3-SECOND SURVEY
The "Automobile Age," particularly in the United States, has revolutionized the way people live, work, and play—and fueled social and economic changes on a huge scale.

3-MINUTE OVERVIEW
The first electric car was the 1888 Flocken Elektrowagen by German inventor Andreas Flocken. Electric cars were popular at the turn of the nineteenth century, but the advantages of gasoline cars—greater range, quicker refueling, and finally the Ford company's mass production (gasoline cars became half the price)—led to a decline. But current concerns about carbon emissions and dwindling fuel resources have reignited interest in hybrid vehicles and electric cars.

RELATED TOPICS
See also
PETROLEUM
page 26

THE WHEEL
page 34

INTERNAL COMBUSTION ENGINE
page 48

3-SECOND BIOGRAPHIES
NICOLAS-JOSEPH CUGNOT
1725–1804
French inventor of first self-propelled vehicle, a steam-power military tractor

KARL FRIEDRICH BENZ
1844–1929
German inventor of first gasoline automobile

EXPERT
Judith Hodge

Ford is credited with producing the first car that gave the middle class the means to travel independently, the Model T, known affectionately as the Tin Lizzie.

THE HELICOPTER

The first helicopter was imagined

by Renaissance artist and inventor Leonardo da Vinci, whose drawings of a "helix" flying machine were inspired by bird flight and the floating maple seed. But the modern helicopter had to wait for advances in aerodynamic theory, structural materials, and more powerful engines in order to get off the ground. Igor Sikorsky's first helicopter, built in 1909, was made of wood and only had a 25 horsepower engine. Thirty years later, he invented the successful VS-300, which went on to become the model for all modern single-rotor helicopters. The invention of the helicopter stems from the evolution of a simple idea: unrestricted flight—moving vertically and horizontally, or hovering in midair. Unlike a fixed-wing plane, the helicopter is capable of vertical lift off and landing—thanks to its rotary blade. It can also hover in a fixed position. These attributes make it ideal wherever space is limited or for anything where it has to stay over a precise area. That means we can use helicopters in anything from rescue work to firefighting, aerial photography and filming, accessing remote locations for environmental or relief work, and to deliver supplies and workers to remote oil rigs—all those uses, despite the fact that helicopters were originally invented solely for military and intelligence purposes.

3-SECOND SURVEY
The invention of the helicopter transformed flight. Its unique design meant it was capable of vertical lift off and had the ability to hover over a single spot.

3-MINUTE OVERVIEW
The helicopter was used by the military almost immediately. Sikorsky's R-4 model was the only helicopter to serve in World War II, where it was used as a direct lift aircraft to rescue people trapped in areas inaccessible by planes. But it was the difficult terrain of the Korean and Vietnam wars where military helicopter use became widespread. Further technological refinements have made it a mainstay of modern warfare.

RELATED TOPIC
See also
THE WRIGHT BROTHERS
page 60

3-SECOND BIOGRAPHIES
LEONARDO DA VINCI
1452–1519
Italian "Renaissance man," gifted artist, mathematician, scientist, and designer-inventor

IGOR SIKORSKY
1889–1972
Russian-American designer of airplanes, flying boats, and the first viable helicopter

EXPERT
Judith Hodge

In spite of the contributions of other engineers to the conception and development of the helicopter, Sikorsky's name is synonymous with the invention.

GLOBAL POSITIONING SYSTEM (GPS)

3-MINUTE OVERVIEW
Due to the potential denial of access and monitoring by the US government, other alternative systems are in use or being developed. These include the Russian Global Navigation Satellite System (GLONASS), which can be added to GPS devices, making more satellites available and enabling positions to be fixed to within 6 feet 6 inches (2 meters). Others include the planned European Union Galileo positioning system, China's BeiDou Navigation Satellite System, and India's Indian Regional Navigation Satellite System (NAVIC).

The global positioning system (GPS) was originally designed for military and intelligence use. The US government developed GPS at the height of the Cold War in the 1960s to track US submarines carrying nuclear missiles. Scientists found that satellites could be tracked from the ground by measuring the frequency of the radio signals they emitted, the so-called "Doppler Effect." GPS uses a network of satellites that orbit Earth and beam down signals carrying a time code and geographical data to a GPS receiver. The invention has transformed navigation, because GPS accurately calculates geographical positions to a matter of yards in all weather conditions, and operates independently of telephone or Internet reception. But it wasn't until 1983, when the former Soviet Union shot down a Korean passenger jet that had strayed into restricted airspace, that the US government opened up GPS for civilian use so that airplanes, shipping, and transport could accurately fix their positions. The US government maintains the system and makes it freely accessible, but it can selectively deny GPS access, for example, to the Indian military in 1999 during hostilities in Kashmir. Today, GPS has many uses beyond navigation such as mapmaking, earthquake research, climate studies, and an outdoor treasure-hunting game known as geocaching.

RELATED TOPIC
See also
THE COMPASS
page 54

3-SECOND BIOGRAPHY
ROGER LEE EASTON
1921–2014
American scientist and physicist, principal inventor and designer of the Global Positioning System, along with Ivan Getting and Bradford Parkinson

EXPERT
Judith Hodge

GPS receivers have been miniaturized to a few integrated circuits. They are now found in cars, planes, and boats, as well as laptop computers.

MEDICINE & HEALTH

adrenaline Hormone produced by the body that increases rates of blood circulation and breathing and makes the heart beat faster, preparing it to react to danger (the "fight or flight response").

anthrax Serious, often fatal bacterial disease of sheep, cattle, and other mammals. It can be transmitted to people via contaminated wool, raw meat, or other animal products, typically affecting the skin and lungs.

antimicrobial resistance Resistance that occurs when microorganisms, such as bacteria and viruses, change in ways that render the medications used against them ineffective.

bacteria Single-cell microscopic organisms found everywhere in Earth's environment; some species are harmless or even beneficial, such as those involved in fermentation or decomposition, while others produce disease-causing toxins.

bite the bullet Expression that originates from a time before effective anesthetics, when soldiers were given bullets to bite on to help them endure pain. Figurative usage means to endure a painful or difficult situation seen as unavoidable.

Boer War Conflicts between the British and the South African Boers; the first (1880–81) when the Boers revolted against the 1877 British annexation of the Transvaal, and the second (1899–1902) fought between the British and the two independent Boer republics of the Orange Free State and Transvaal.

chicken pox Highly contagious disease most common in children, caused by herpes virus and characterized by itchy, inflamed blisters that eventually scab over.

chloroform Coloulous, volatile liquid with a sweet taste and odor, used as a solvent and in refrigerants; formerly used as an anesthetic.

cowpox Contagious skin disease of cattle, usually affecting the udder, caused by a virus; resembles mild smallpox when contracted by humans through contact, and was the basis of the first smallpox vaccines.

diabetes Disease in which the body's ability to produce or respond to the hormone insulin is impaired, resulting in elevated levels of glucose in the blood and possible health complications, including heart disease and blindness.

diphtheria, tetanus, and pertussis (DTP) immunization Combined vaccine to immunize children. Its side effects led to the development of safer acellular vaccines (DTaP).

electromagnetic radiation Form of energy associated with electric and magnetic fields that includes visible light, radio waves, microwaves, gamma rays, and X-rays.

ether Highly flammable, colorless liquid used as a solvent in industrial processes; formerly used as an anesthetic.

hepatitis Inflammation of the liver, most commonly caused by hepatitis virus; there are several types, including hepatitis A, typically caused by contaminated food or water, and hepatitis B, transmitted by infected body fluids such as blood.

HIV Human immunodeficiency virus, which damages the immune system; if untreated, HIV can result in the potentially life-threatening condition AIDS (acquired immunodeficiency syndrome).

malaria Serious infectious disease of humans (and other animals) spread by certain mosquitoes, characterized by recurrent symptoms of chills, fever, and an enlarged spleen; common in tropical climates and endemic in many developing countries. There is currently no vaccine available.

measles, mumps, and rubella (MMR) vaccine Combined vaccine to immunize children; safety concerns involving the vaccine, including a recent controversy over possible links to autism, have been disproved.

meningitis Disease involving inflammation of the membranes (meninges) surrounding the brain and spinal cord, caused by viral or bacterial infection.

nitrous oxide Colorless, odorless gas used in surgery and dentistry as a mild anesthetic; known as "laughing gas" due to the euphoric effects of inhaling it, a property that has led to its recreational use.

rabies Contagious and potentially fatal viral disease of dogs and other mammals, transmissible through the saliva to humans, usually by the bite of infected animals; it can cause madness and convulsions.

streptococcus Group of bacteria that causes illnesses, such as strep pneumonia, strep throat, scarlet fever, and rheumatic fever.

thimerosa Mercury-containing preservative to prevent growth of bacteria or fungi; no longer used in vaccines for babies and young children due to safety concerns.

tuberculosis (TB) Infectious disease caused by bacteria, characterized by the growth of nodules mainly in the lungs; it can be treated, but drug-resistant TB is increasing.

VACCINATION

3-SECOND SURVEY
Vaccination may have eradicated diseases, such as smallpox, although vaccines for other "plagues," such as malaria and HIV, prove more elusive.

3-MINUTE OVERVIEW
The 1956 WHO vaccination campaign to eradicate smallpox was declared effective in 1980, after the last naturally occurring case of smallpox was reported in Somalia in 1977. Viruses, such as polio, have remained more resistant to vaccination efforts, although the disease is now rare. Controversy has surrounded the safety and efficacy of the diphtheria, tetanus, and pertussis (DTP) immunization; the measles, mumps, and rubella (MMR) vaccine; and the use of a mercury-containing preservative called thimerosal.

Vaccines are often described as one of the greatest achievements of public health. It is estimated that widespread vaccination has reduced morbidity by 99 percent for five "killer" diseases: smallpox, diphtheria, measles, polio, and rubella. Smallpox was a particular scourge for centuries. A basic form of vaccination, known as variolation, was practised in China and India as early as 900 CE where healthy people were exposed to powdered scabs from smallpox pustules inserted either under the skin or in the nostrils. An English doctor, Edward Jenner, is generally recognized as the inventor of the first "modern" vaccine. Jenner observed that people who had suffered from the less virulent cowpox seemed immune to smallpox. In 1796, he carried out his famous experiment on eight-year-old James Phipps, inserting pus from a cowpox pustule into the boy's arm. Neither James or the other children he vaccinated developed smallpox (Jenner took the name vaccine from the Latin *vacca* for "cow"). Despite opposition from both the medical profession and the Church, the invention was widely adopted in Europe and the United States, and further developed to protect against other diseases. Louis Pasteur developed vaccines for anthrax and rabies in the nineteenth century, using "killed" agents rather than live viruses.

RELATED TOPICS
See also
ANTIBIOTICS
page 76

THE MICROSCOPE
page 78

LOUIS PASTEUR
page 80

3-SECOND BIOGRAPHIES
EDWARD JENNER
1749–1823
English doctor who invented first "modern" vaccine for smallpox

LOUIS PASTEUR
1822–95
French microbiologist made famous for his discoveries of the principles of vaccination and pasteurization

MAURICE HILLEMAN
1919–2005
American microbiologist and most prolific vaccine inventor

EXPERT
Judith Hodge

Vaccinations have been widely available since the 1920s; many of those currently licensed were invented by Hilleman.

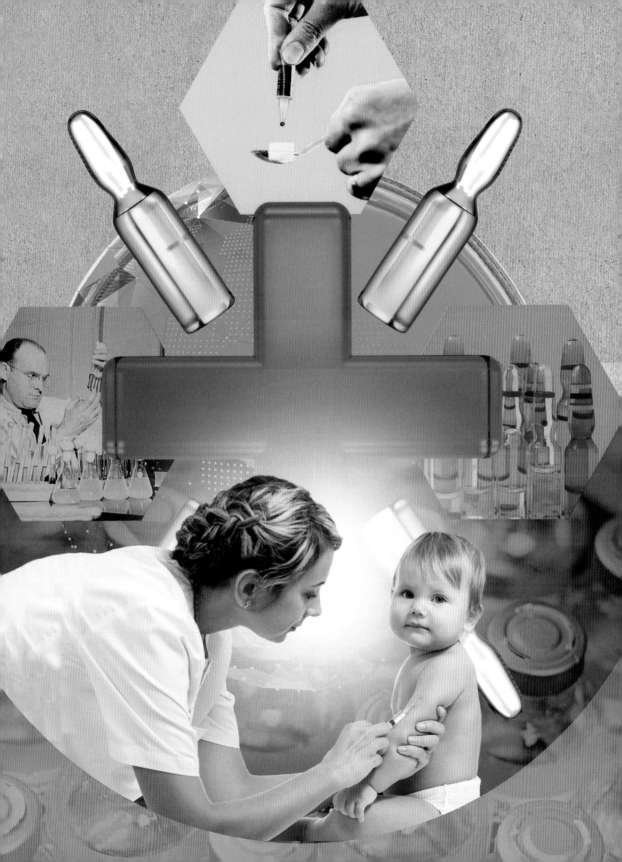

ANESTHESIA

The search for an effective method to control pain, particularly for surgical operations, took centuries to resolve. Alcoholic concoctions were used as early anesthetics, as were plants containing powerful chemicals. A breakthrough came in 1846, when American dentist William Morton instructed a patient to inhale the fumes of an ether-soaked sponge in front of physicians and students at the Massachusetts general hospital. Once the patient was sedated, the surgeon, John Warren, removed a tumor from the patient's neck. When he awoke, the patient reportedly declared that "it feels as though my neck's been scratched." Up until then, surgery was quick and brutal, with fully conscious patients having to bite the bullet—sometimes literally—during the operation. Although ether had been known of for about 300 years, no one had thought to use it as an anesthetic. It had instead been used as a recreational drug at parties known as "ether frolics." News of Morton's success quickly spread and numerous operations using ether were performed by respected European surgeons. Anesthetic obviously made surgery a better experience for patients, but it also gave surgeons more time when operating, which in turn has led to advances in surgical techniques. In time, ether was replaced by safer and more reliable drugs.

3-SECOND SURVEY
Anesthesia gives temporary relief from pain so that surgeons can perform their work quickly without their patients dying from shock.

3-MINUTE OVERVIEW
Modern chemistry led to the development of isolated anesthetics. Of these early drugs, only nitrous oxide (laughing gas) is still widely used today. Both ether (which is highly flammable and causes excessive vomiting) and chloroform (which is pharmacologically dangerous) have been replaced by safer general anesthetics. Current anesthesia practices involve the use of one or more drugs tailored to the particular patient and procedure, although there are still major risks, such as heart attack or pulmonary embolism.

RELATED TOPICS
See also
VACCINATION
page 72

ANTIBIOTICS
page 76

3-SECOND BIOGRAPHIES
WILLIAM MORTON
1819–68
American dentist who first demonstrated the use of ether as a surgical anesthetic in 1846

JOHN SNOW
1813–58
English physician who popularized obstetric anesthesia by chloroforming Queen Victoria for the birth of Prince Leopold, her eighth child, in 1853

EXPERT
Judith Hodge

The field of anesthetics has developed into a highly skilled medical practice since the experiments made by early practitioners, such as William Morton and John Snow.

ANTIBIOTICS

3-SECOND SURVEY

The advent of antibiotics hailed a new era in the treatment of communicable disease, because it meant that people no longer died from common infections.

3-MINUTE OVERVIEW

The effectiveness and accessibility of antibiotics has led to their overuse, particularly as growth promoters in animals and fish. Microbial resistance results in "superbugs" that no longer respond to antibiotics, now seen in forms of tuberculosis and increasingly in HIV and malaria. In 2014, the World Health Organization (WHO) classified antimicrobial resistance as a "serious threat" affecting every region of the world. Researchers are currently exploring alternatives to antibiotics, including predatory bacteria and metals.

Antibiotics revolutionized medicine in the twentieth century. Before 1935, people routinely died from minor cuts and scratches that had become infected. Penicillin, discovered in 1928, is often cited as the first antibiotic, but drugs containing sulfur were being used to treat bacterial infections a decade earlier. The bright orange-red compound prontosil was used as an industrial dye before the German scientist Gerhard Domagk started testing it on mice injected with a lethal dose of *streptococcus*. The mice given prontosil survived. The biggest test for Domagk came when his young daughter developed a *streptococcus* infection in 1935; despite minimal testing on humans, he made the desperate decision to give her large oral doses of prontosil and, two days later, she was miraculously cured. Domagk was given the Nobel Prize for his discovery in 1939. But penicillin is still the world's most famous antibiotic. Scottish biologist Alexander Fleming noticed that mold had developed on dishes being used to grow the *staphylococci* germ, and had created a bacteria-free circle around itself. But he needed to do more work to develop penicillin into a mass-produced medicine. Penicillin was in big demand during World War II, when it was widely used to treat infected wounds and diseases, such as syphilis.

RELATED TOPICS

See also
VACCINATION
page 72

ANESTHESIA
page 74

3-SECOND BIOGRAPHIES

ALEXANDER FLEMING
1881–1955
Scottish biologist whose accidental discovery of penicillin is widely heralded as the start of modern antibiotics

GERHARD DOMAGK
1895–1964
German bacteriologist credited with creating the first commercial antibiotic

EXPERT

Judith Hodge

Hailed as one of the "wonder drugs," of the last century, few new families of antibiotics have been created since the early 1960s. Fleming himself cautioned against penicillin misuse.

THE MICROSCOPE

The Dutch scientist Anton van Leeuwenhoek is credited with being the first person to make and use a practical microscope. Before his late seventeenth-century innovations, the ancient Romans and Greeks and the Chinese had used lenses for magnification. In the 1590s, two Dutch eyeglass-makers, the Janssens, put several lenses in a tube and discovered that an image magnified by a single lens can be increasingly magnified by a second or further lenses. With a "compound" microscope like this, the English scientist Robert Hooke made observations and drawings of the hairs on the bodies of fleas, captured in his 1665 book *Micrographia*. He also discovered pores in thin slices of cork, calling them plant "cells," because they reminded him of cells in a monastery. Leeuwenhoek achieved greater success with magnification by developing new ways of grinding lenses to improve the optical quality—up to 200x magnifying power when other microscopes of the time had 20–30x magnification. His handheld microscope used a single convex glass lens attached to a metal holder and was focused using screws. In a famous experiment in 1674, he viewed the teeming, previously unseen life in a drop of lake water. He was the first person to describe bacteria, yeast plants, and the circulation of blood corpuscles in capillaries.

3-SECOND SURVEY
Before the microscope, people did not realize that living things were made up of tiny structures that could not be seen with the naked eye.

3-MINUTE OVERVIEW
Optical microscopes are only able to focus on objects that are at least the size of a wavelength of light. The invention of the electron microscope in 1931 by Max Knoll and Ernst Ruska shattered the limitations of visible light by transmitting a beam of electrons (instead of light) through a specimen. Ruska's principles still form the basis of modern electron microscopes, which are able to achieve magnification levels of up to two million times.

RELATED TOPICS
See also
THE OPTICAL LENS
page 82

THE X-RAY
page 84

3-SECOND BIOGRAPHIES
ZACHARIAS JANSEN
c. 1580–c. 1638
Dutch eyeglass-maker, who, with his father Hans, invented the first microscope

ANTON VAN LEEUWENHOEK
1632–1723
Dutch draper and scientist, "father" of the microscope

ERNST RUSKA
1906–88
German physicist who won the Nobel prize in 1986 for his work on electron optics, including the invention of the electron microscope

EXPERT
Judith Hodge

Great advances in microscopy had been made by the mid-1600s, after which microscopes remained unchanged for the next 200 years.

December 27, 1822
Born in Arbois, France, third child of a poor tanner Jean-Joseph Pasteur, and Jeanne-Etiennette Roqui

1848
Discovers "isomerism," existence of left- and right-handed molecules as chemist at University of Strasbourg

May 29, 1849
Marries Marie Laurent: they have five children, only two of whom survive to adulthood

1859
Develops controversial "germ theory," through experiments with fermentation

1862
Wins prize for disproving theory of spontaneous generation; invents pasteurization

1865–67
Rescues European silk industry through discovery of microbes attacking silkworm eggs

1868
Has a stroke at the age of 45

1879
Works on chicken cholera, and man-made vaccines

1881
Develops vaccine for anthrax

July 6, 1885
Vaccinates a nine-year-old boy, Joseph Meister, against rabies

1887
Founds the Pasteur Institute; remains director until his death

September 28, 1895
Dies at 72 in Marnes-la-Coquette from another stroke

LOUIS PASTEUR

As a child, Louis Pasteur showed more interest in painting and drawing than science. His first discovery, as a chemist at the University of Strasbourg, was that almost identical molecules could exist as left- and right-handed "mirror" versions. This had huge implications for microbiology (of which Pasteur is called the "father") and drug development.

In 1849, he married Marie Laurent, who assisted him in all his scientific experiments. Only two of their five children survived to adulthood, the others dying of typhoid. Pasteur wrote that such personal tragedies motivated him to find cures for infectious diseases.

A breakthrough came when he developed the controversial "germ theory." In 1862, the French Academy of Sciences offered a prize of 2,500 francs to anyone who could prove or disprove that life was spontaneously generated. In his most famous experiment, Pasteur demonstrated that food was contaminated due to microbes in the air but could be sterilized by heating. He achieved fame with his discovery when the theory was put into practice.

France's wine trade was suffering huge losses from bottles spoiled in transit. The invention of pasteurization—in this case, briefly heating wine to 131°F (55°C)—killed off microorganisms without affecting the taste. The process was extended to other farm and brewery products such as milk and beer. Beverage contamination led Pasteur to the idea that microorganisms in animals and humans might also be causing disease, and his research inspired British surgeon Joseph Lister to develop antiseptic methods in surgery.

Despite not being a biologist, Pasteur took up the challenge to save Europe's silk industry from an unknown disease. By 1868, he had discovered that not one but two parasitic microbes were attacking the worms' eggs, and that disinfecting the nurseries would solve the problem.

Pasteur had a stroke at 45, probably due to overwork, but a mobile laboratory was set up at his bedside. Through his study of chicken cholera he built on Edward Jenner's earlier work, and manufactured vaccines for both rabies and anthrax. However, recent studies of his notebooks reveal that Pasteur may have overstated how much original work he did. He secretly used the method of veterinarian Jean Toussaint to prepare an anthrax vaccine. The first human trial of a rabies vaccine, when Pasteur vaccinated a boy who had been bitten by a rabid dog, was a success, but he had exaggerated how much animal testing he had carried out beforehand.

In 1887, Pasteur founded the research institute that bears his name. Although he achieved many things, Pasteur is probably best remembered for the invention that commemorates his name, pasteurization.

Judith Hodge

THE OPTICAL LENS

3-SECOND SURVEY

The invention of the optical lens greatly prolonged productivity, because people were able to overcome the natural aging process of the human eye.

3-MINUTE OVERVIEW

New optical instruments, such as eyeglasses, the microscope, and the telescope, had a huge impact on sixteenth- and seventeenth-century society and its view of the world. Galileo Galilei's new, improved telescope design enabled him to see craters on the Moon and four moons orbiting Jupiter. With his "terrestrial spyglass," he developed his theory that the Earth moved around the Sun—an idea that was considered heresy at the time.

The first lenses, dated as early as 700 BCE, were polished rock crystals. The Roman emperor Nero is said to have watched gladiators using an emerald as a corrective lens. Two early Arabic scientists transformed how light and vision were understood. Ibn el-Haitam was the first to correctly describe how light is refracted by a lens (and how the eye functions); previously people had believed that light was released rather than reflected by objects. Ibn al-Haytham (Alhazen) wrote about the effects of concave lenses and magnifying glasses in his *Book of Optics*. The book was translated into Latin in the twelfth century and inspired Franciscan monk Roger Bacon to recommend glass spheres as magnifying glasses to help people read. By the thirteenth century, Venice and Florence were home to an optical industry, grinding and polishing lenses. It was in Italy that the first wearable eyeglasses (using lenses for both eyes) were invented at the end of the 1280s. Literacy was not widespread in the fourteenth century, but the invention of printing led to a huge increase in books and pamphlets, and a corresponding rise in reading and need for reading glasses. Further development of the optical lens by Dutch eyeglass-makers led to the invention of the microscope and the refracting telescope in 1608.

RELATED TOPICS

See also
THE MICROSCOPE
page 78

THE PRINTING PRESS
page 98

3-SECOND BIOGRAPHIES

IBN AL-HAYTHAM (ALHAZEN)
c. 965–c. 1040
Arabic scientist, author of influential *Book of Optics*

ROGER BACON
c. 1214–94
English Franciscan monk and philosopher whose study of the optical properties of lenses and mirrors led to charges of "witchcraft" against him

GALILEO GALILEI
1564–1642
Italian polymath who developed the Galilean telescope and mapped the night sky

EXPERT

Judith Hodge

Eyeglasses with optical lenses were first worn in 1284 in Italy, replacing "reading stones" made by cutting a glass sphere in half.

THE X-RAY

The invention of the X-ray in 1895 meant that for the first time the inner workings of the body could be made visible without surgery. While studying the effects of passing an electrical current through gases at low pressure, German physicist Wilhelm Röntgen accidentally discovered that electromagnetic radiation was capable of penetrating most solid objects. He named these X-rays ("X" stood for unknown), and later sent an X-ray photograph of his wife's hand—with bones and a wedding ring clearly visible—to colleagues. The discovery transformed medicine almost overnight. Within a year, the first radiology department had opened at Glasgow Hospital, producing the first pictures of a kidney stone and a penny lodged in a child's throat. Shortly after, X-rays were used to trace the path of food through the digestive system. Röntgen's discovery was used in the Boer War and in World War I, to locate bone fractures and embedded bullets in wounded soldiers. There was much public excitement at the new technology, and X-ray machines appeared as a wondrous curiosity in theatrical shows. Today, X-rays are most widely used in medicine to analyze problems with bones, teeth, and organs in the body. They also have uses in industry for tasks, such as detecting cracks in metal and security checks at airports.

3-SECOND SURVEY

Röntgen's accidental discovery of the X-ray rapidly transformed medicine by enabling people to see the inside of the human body without cutting into flesh.

3-MINUTE OVERVIEW

The widespread and unrestrained use of X-rays led to serious injuries, with people who were X-rayed or who worked with the original machines suffering radiation burns, hair loss and organ damage. Early dosages were up to 1,500 times higher than those used today. It was eventually recognized that frequent exposure to X-rays could be harmful. Today, special measures are taken to protect the patient and physician. High doses of X-rays are used to treat cancer.

RELATED TOPICS

See also
ANESTHESIA
page 74

ANTIBIOTICS
page 76

THE MICROSCOPE
page 78

3-SECOND BIOGRAPHY

WILHELM RÖNTGEN
1845–1923
German physicist awarded the first Nobel Prize in physics in 1901 for his discovery of X-rays, an invention that continues to be used in hospitals worldwide

EXPERT

Judith Hodge

X-rays are used in other branches of science, such as crystallography, which was used by Rosalind Franklin to discover the structure of DNA, and microscopic analysis, which uses X-rays to produce images of very small objects.

THE BIRTH CONTROL PILL

3-SECOND SURVEY
The pill was one of the most significant medical advances of the twentieth century, because it changed the way that sexual relationships were viewed.

3-MINUTE OVERVIEW
Health scares have dogged the pill from the beginning, with reports linking its use with blood clots, strokes, diabetes, and heart attacks. Research in the 1980s suggested links between pill use and breast cancer, causing user numbers to dip. Some of the concerns were linked to hormone levels in the pill, which have now been lowered. On the plus side, the pill has been shown to be protective for ovarian cancer and pelvic inflammatory disease.

Inextricably linked to free love, the "swinging sixties," and women's liberation, the contraceptive pill was developed in the 1950s by American biologists Gregory Pincus and John Rock. Arguably, few inventions have had such a profound social impact in terms of transforming women's lives—though the pill's road to widespread use has been a rocky one. At a New York dinner party, birth control activist Margaret Sanger persuaded Pincus to work on a contraceptive pill, a combination of the hormones oestrogen and progesterone. In 1956 large-scale clinical trials conducted in Puerto Rico deemed the pill to be 100 percent effective, but some serious side effects were ignored. The following year the Federal Drug Administration (FDA) approved the pill, but only for severe menstrual disorders—an unusually large number of women reported severe menstrual disorders. When it was approved for contraceptive use in 1960, take-up was rapid: within five years, it was the most popular form of birth control in the United States, used by 6.5 million American women. It was introduced in the UK on the NHS in 1961, only for married women—this lasted until 1967—and it is now taken by 3.5 million women in Britain between the ages of 16 and 49. Currently the pill is taken by about 100 million women worldwide.

RELATED TOPICS
See also
VACCINATION
page 72

ANTIBIOTICS
page 76

3-SECOND BIOGRAPHIES
MARGARET SANGER
1879–1966
American birth control activist who opened the first birth control clinic and started the Planned Parenthood Federation of America, which funded the pill's development

JOHN ROCK & GREGORY PINCUS
1890–1984 & 1903–67
American coinventors of the pill

CARL DJERASSI
1923–2015
Bulgarian-American chemist who chemically created the pill by synthesizing Mexican yams in 1951 but was unable to manufacture or distribute it

EXPERT
Judith Hodge

The pill was the first lifestyle medicine; it gave women control over their own fertility.

THE PACEMAKER

The invention of the pacemaker, a device that emits electrical pulses that prompt the heart to beat at a normal rate, was literally lifesaving. Before its advent, there were only two methods to help someone with irregular heart rhythms or experiencing cardiac arrest: mechanical stimulation of the heart or injecting a medicine, such as adrenaline, through the chest wall. Both were unreliable. Early pacemakers were external machines as big as TV sets and gave electric shocks when used. Albert Hyman's 1932 invention was powered by a hand-cranked motor. Public perception that pacemakers were interfering with nature by "reviving the dead" may have curtailed research until the 1950s. In 1958, Arne Larsson became the first person to receive an implantable pacemaker attached by electrodes to the heart's muscle tissue. The device was designed by Swedish physician and inventor Rune Elmqvist, working closely with cardiac surgeon Åke Senning. The patient went on to receive 26 different pacemakers during his lifetime and died at 86, having outlived both inventor and surgeon. The next important innovation was overcoming the short lifetime of early pacemaker batteries. In 1971, Wilson Greatbatch invented the long-life lithium-iodine battery, which became the standard for future pacemaker designs.

3-SECOND SURVEY
The invention of the pacemaker has saved thousands of lives, using electrical pulses to regulate the beating of the heart.

3-MINUTE OVERVIEW
Modern pacemakers have external programming tailored for individual patients so that a cardiologist can select the optimum pacing modes. Some combine a pacemaker and defibrillator in a single implantable device. Others have multiple electrodes stimulating differing positions within the heart to improve the way the lower chambers (ventricles) of the heart are synchronized. The latest pacemaker technology is a new wireless device about the size of a grain of rice, inserted directly into the heart.

RELATED TOPIC
See also
THE X-RAY
page 84

3-SECOND BIOGRAPHIES
ALBERT HYMAN
1893–1972
American cardiologist who, along with brother Charles, built one of the earliest "artificial pacemakers"—he was the first to use the term

RUNE ELMQVIST
1906–96
Swedish physician, engineer, and inventor who designed the first implantable pacemaker

WILSON GREATBATCH
1919–2011
American engineer, inventor, and holder of 350 patents, including the lithium-iodine battery used as standard cell for pacemakers

EXPERT
Judith Hodge

By the mid-1980s, pacemakers—now the size of a large coin—were routinely fitted to people with heart conditions.

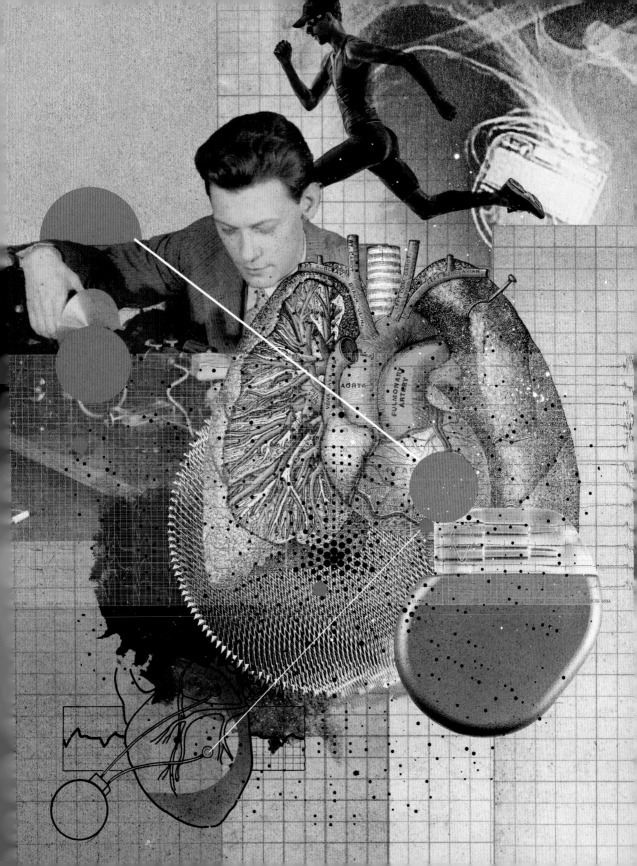

COMMUNICATIONS

COMMUNICATIONS
GLOSSARY

abugida (abjad) Prealphabetic system, also known as consonantal; such systems have no symbols representing vowels.

alternator Invented by Ernst Alexanderson in 1904, a rotating machine for generating the continuous radio waves necessary for the transmission of amplitude modulation (AM) sound, making the broadcasting of speech possible.

Analytical engine Charles Babbage's 1837 proposition for the first general mechanical computer. It contained an Arithmetic Logic Unit, basic flow control using punch cards modeled on the Jacquard loom, and integrated memory. Like Babbage's difference engine, it was not built during his lifetime.

audion Electron tube consisting of three electrodes, cathode filament, anode plate, and control grid mounted in a glass or metal vacuum tube, used to amplify radio signals.

binary maths Numeral system that writes numbers using only two digits: 0 and 1. These are used in computers as a series of "off" and "on" switches. In binary, each digit's place value is twice as much as that of the next digit to the right (because each digit holds two values).

Boolean logic Developed by George Boole, a mathematician and computer pioneer, a Boolean, or bool, consists of operators, such as AND, OR, NOT, and XOR. Booleans are often used in programming and in Internet search engines. Boolean expressions are ones that result in the value of either TRUE or FALSE.

cathode ray Beam of electrons in a vacuum tube traveling from the negatively charged electrode (cathode) at one end to the positively charged electrode (anode) at the other, across a voltage difference between the electrodes. Used for generating pictures in traditional television and computer monitors.

coherer Early form of radio detector consisting of a glass tube loosely filled with metal filings whose bulk electrical resistance decreased in the presence of radio waves. The identity of the inventor of the coherer remains disputed. It was patented by Marconi in 1901 but seems to have originated with the work of Jagadish Chandra Bose.

Colossus First electric programmable computer, developed by English engineer Tommy Flowers and built in 1943. It was used by the British during World War II to read encrypted German messages.

galley Tray into which assembled type was placed prior to printing.

hieroglyphs Pictographic writing system of the ancient Egyptians (3200 BCE–400 CE).

HTML Hypertext markup language, how web pages are published.

HTTP Hypertext transfer protocol—serves up web pages on request.

Kinescope tube Name given to first incarnation of electronic television, invented by Vladimir Zworykin in 1929.

klystron High-frequency amplifier for generating microwaves, invented by brothers Russell and Sigurd Varian, the technology that made ultra high frequency (UHF) television possible.

lingua franca Language used in common between speakers of diverse native languages.

mechanical television Earliest attempts to produce television, based on perforated rotating disks and the use of mirrors.

Morse code System for sending messages invented by Samuel Morse and Alfred Vail in the late 1830s. Morse code assigns each letter and number a unique set of dots or dashes.

movable type machine Technology for printing that uses separate and reusable characters made from metal.

Orthicon Camera developed by Vladimir Zworykin in 1943 that had sufficient light sensitivity to record outdoor events at night.

OS Short for operating system, the most important software that runs on a smartphone (or computer). It manages the phone's memory, processes, and all its software. It lets us communicate with the phone without knowing how to speak its language.

phonemic alphabet Abstract writing system in which each symbol represents a sound.

Proto-Sinaitic script First alphabet, adapted from Egyptian hieroglyphs by the peoples of Egypt and Sinai.

thalassocracy From the Greek *thalassa* meaning "sea," and *kratein*, meaning "to rule"; a state whose power derives from naval or maritime supremacy.

URL Uniform resources locator, a universal system for recognizing the location of web pages.

THE ALPHABET

The ancient Greeks created the

first true alphabet in the eighth century BCE. Their breakthrough innovation was to use characters to represent vowels, completing a process that had begun thousands of years earlier, and producing a writing system that could record speech without ambiguity and was accessible to ordinary people. Archaeological evidence suggests that, by around 1700 BCE, Semitic workers had already simplified the alphabet from the unwieldy hieroglyphs of the ancient Egyptians. Reducing the number of characters to a couple of dozen symbols, "Proto-Sinaitic" script was useful for commerce, which led to it becoming widespread, with many variants created. The Phoenician thalassocracy of the eighth century BCE created the phonemic alphabet by adapting characters to represent consonants instead of syllables. Trading throughout Asia Minor and the Mediterranean, the Phoenicians used their consonantal alphabet to record any language they encountered. Because the transactions they recorded related to money, there was huge motivation among their trading partners to learn Phoenician script, thereby spreading its use. But the lack of vowels was a source of confusion. The Greeks solved this problem by repurposing characters for which there was no equivalent sound in the Greek language to represent vowels.

3-SECOND SURVEY
The most widely used alphabets of the modern world include Greek, Cyrillic, Arabic, and Latin, all of which have an ancient common ancestor in the Near East.

3-MINUTE OVERVIEW
The creation of the alphabet was prompted by the needs of trade and exploration among widely disparate peoples. Commerce, diplomacy, and even war required the accurate recording and transmission of information. The simplification achieved by the adoption of consonants with a reduced number of abstract symbols, and the subsequent inclusion of vowels, increased accessibility and provided an environment that allowed for the development of codified law, logic, science and history.

RELATED TOPICS
See also
THE PRINTING PRESS
page 96

THE COMPUTER
page 98

THE TELEGRAPH
page 100

3-SECOND BIOGRAPHY
PIERRE MONTET
1855–1966
French explorer who in 1923 discovered the sarcophagus of King Ahiram with its inscription in Old Phoenician alphabet in Jbeil, Lebanon

EXPERT
Diana Rawlinson

Any alphabet provides the means to write one or more languages using letters. Some early abudgida (alphabets without vowels) are still in evidence today in Arabic and Hebrew.

THE PRINTING PRESS

3-SECOND SURVEY
The invention of the printing press introduced books to the masses as well as heralded the demise of Latin as the "lingua franca."

3-MINUTE OVERVIEW
Four centuries before Gutenberg, Chinese artisan Pi Sheng was composing texts using characters molded from a mixture of clay and glue set on an iron plate. Due to the complexity of the Chinese alphabet, which was primarily pictographic not alphabetic, and the lack of press technology at the time —the Chinese made neither wine nor olive oil!—printing presses did not take off in China at this time.

Johannes Gutenberg was the first person to demonstrate the practicability and usefulness of a "movable type machine" around 1440 in Mainz, Germany. A goldsmith by trade, he was quick to recognize the market potential of the mass production of books as literacy among the lay people of Europe spread during the Renaissance. Gutenberg's innovation combined individual letters uniform in size and cast from a metal alloy of lead, tin, and antimony, a modified screw press, and oil-base inks to produce a durable and reusable technology. Each individual letter of the alphabet could be moved around to form words, words were separated by blank spaces to form lines of print, and those lines were brought together in a galley to make up a page. The capacity to reset the type in any format made it possible to create an infinite variety of texts. The immediate effect of the printing press was to increase the availability of books and make them cheaper to acquire. This in turn drove the dissemination and standardization of information, crucial to the development of education, science, and technology. Despite Gutenberg's efforts to keep his techniques secret, more than 250 European cities had printing presses by 1500, producing an estimated 20 million volumes. His most famous printing, some 180 copies of the Bible, sold out at the Frankfurt Book Fair of 1455. Remarkably, around 50 survive today.

RELATED TOPICS
See also
TIM BERNERS-LEE
page 102

THE SMARTPHONE
page 108

3-SECOND BIOGRAPHIES
PI SHENG
990–1051
Chinese artisan who invented the first movable type by carving characters into clay

JOHANNES GUTENBERG
1398–1468
German goldsmith and stonecutter who invented a "movable type machine"

ALDUS PIUS MANUTIUS
1449–1515
Italian humanist who founded the Aldine Press in Venice in 1494, producing respected editions of the classics

EXPERT
Diana Rawlinson

Gutenberg's Bible was the world's first substantial work to be printed, instead of being hand copied.

THE COMPUTER

Before the Industrial Revolution, machines were designed to do one thing—for example, to tell the time or grind corn. But greater mechanization called for something more adaptable. In 1801, Joseph Jacquard built a loom in which interchangeable punch cards were used to control the patterns of cloth produced—the first machine that could be instructed to do more than one task. It was the inspiration for Charles Babbage's "Difference Engine," conceived in 1822. His idea was for a universal machine capable of limitless calculations that would minimize error and analyze the accuracy of the results. Financial and technological limitations meant that Babbage's machine was never built in his lifetime, but perhaps the reason it took until the twentieth century for the first programmable computers to be developed was that the demand didn't exist. Yet, as engineering and military needs became more complex, particularly during World War II, issues requiring computations beyond the capabilities of humans in scale and time created that demand. When Alan Turing published his hypothesis of a Turing Machine in 1936 (defined as a machine that can perform any computation that could be performed by any other computing device), he formalized the basis of computing, which together with the application of electronics, provided the foundation for modern computers.

RELATED TOPICS

See also
THE TELEVISION
page 106

THE SMARTPHONE
page 108

3-SECOND BIOGRAPHIES
ADA LOVELACE
1815–52
English mathematician who wrote an algorithm for use by Babbage's analytical engine

KONRAD ZUSE
1910–95
German inventor who built the first fully program-controlled computer, the Z1, in 1938 and the first functional programmable computer, the Z3, in 1941

ALAN TURING
1912–54
English mathematician whose analysis of methodical process led to the development of "definite method"—in modern language, an algorithm

EXPERT
Diana Rawlinson

ENIAC, the world's first electronic computer, cost $500,000 and occupied the same floor space as a small house.

THE TELEGRAPH

3-SECOND SURVEY

Instantaneous communication through use of the telegraph changed the way people thought about how the world was organized.

3-MINUTE OVERVIEW

The electric telegraph has been called "the Victorian Internet" because of the similar effects that both inventions had on society and the mixed reception they received. Both use coding to convey messages, creating divisions between those adept at manipulating them and those who only receive the message. Both are subject to concerns about security and notions of information sensitivity, and both transformed connectivity between people, with the potential to unite and to divide.

In the early nineteenth century, scientists were searching for a communication system that could cross large distances without relying on lines of sight. The telegraph emerged when they found they were able to use new developments in the study and understanding of electricity. In Britain, William Fothergill Cooke and Charles Wheatstone produced an electric telegraph that was used for railroad signaling, while in the United States, Samuel Morse, Alfred Vail, and Leonard Gale developed a battery-power system needing only a wire to connect to the receiver and a key to tap out the dots and dashes of Morse code. Winning government money to build a telegraph line from Washington, DC, to Baltimore, Morse and Vail sent their first 19-letter message in May 1844. It read: "What hath God wrought." The potential of the telegraph to speed up communications became obvious to commercial interests, which contributed to the rapid expansion of wire connections across the country. The impact was to increase the volume of commerce; markets could be more efficiently exploited and information more easily transferred. This in turn led to more centralized business and administrative systems. The electric telegraph heralded a revolution in communication that highlighted the value of knowledge itself and changed the face of diplomatic and military affairs.

RELATED TOPICS

See also
THE COMPUTER
page 98

THE RADIO
page 104

THE TELEVISION
page 106

3-SECOND BIOGRAPHIES

ALESSANDRO GIUSEPPE VOLTA
1745–1827
Italian physicist and chemist who developed the first battery and discovered methane gas

SAMUEL FINLEY BREESE MORSE
1791–1872
American portrait painter and inventor of the electric telegraph, who coinvented Morse code with Alfred Vail

EXPERT
Diana Rawlinson

The electric telegraph marked the beginning of experimentation with unseen forces, using energy itself.

June 8, 1955
Born in London, England, to Conway Berners-Lee and Mary Lee Woods

1976
Awarded a first-class degree in physics at the Queen's College, Oxford

1980
While working at CERN, he first proposes the concept of using "hypertext" to facilitate the sharing of information

1989
Produces "Information Management: A Proposal"—a system for sharing and distributing information globally; he calls it the World Wide Web

1991
Creates the first web browser and editor, and launches the first ever website, built at CERN "info.cern.ch." It explains the World Wide Web and gave users an introduction to getting started with their own websites

1994
Founds the world wide web (W3C) consortium committed to open web standards at the Laboratory of Computer Science at the Massachusetts Institute of Technology

1999
Named one of the 100 Most Important People of the Twentieth Century by *Time* Magazine

2004
Knighted and awarded the Finland Millennium Prize

2007
Awarded the UK's Order of Merit, a personal gift of the monarch, and limited to 24 living individuals.

2009
Elected a fellow of the American Academy of Arts and Sciences

2011
One of the three recipients of the Mikhail Gorbachev award for The Man Who Changed the World

2012
Co-founds the Open Data Institute advocating for open data globally

2012
Plays a starring role in the opening ceremony of the London Olympics viewed by 900 million people; he tweets "this is for everyone"

TIM BERNERS-LEE

Tim Berners-Lee's fascination with things electronic began when he was a child, creating gadgets to control his model trains. Both his parents were computer scientists and worked on one of the first commercially built computers, the Ferranti Mark I, so his career choice was perhaps no surprise.

After graduating from Oxford with a degree in physics, Berners-Lee worked as a software engineer and consultant, including at CERN, the particle physics laboratory in Geneva. It was here that he first addressed the problem of information sharing with a visionary approach that grew to become the World Wide Web.

The "Internet," simply meaning the linking of one computer to another, had begun in the 1960s among researchers and academics as a way of transferring content. The limitation that Berners-Lee identified at CERN was that each research team collected their data on a different network of computers, sometimes using a different programming language, which complicated participation and collaboration. His solution was to combine the internet with an emerging technology called hypertext, a system that allowed onscreen viewing of documents from anywhere on the network, without having to download them. His prototype software was called ENQUIRE.

Berners-Lee's 1989 proposal for the World Wide Web was not taken up as an official project at CERN, but he continued to work on it, and by 1990, together with informatics engineer Robert Cailliau, he had written the three fundamental technologies that remain the foundation of the web: HTML (hypertext markup language), URL (uniform resource locator) and HTTP (hypertext transfer protocol), and produced the first web browser and the first web server.

From the beginning, Berners-Lee realized that the true potential of the web could only be exploited if the underlying code was available royalty-free so that anyone could use and develop it, and he has continued to advocate for open data, net neutrality, and for the web to serve the global public good. Beyond his technical contribution, this commitment to open resources is his most significant, and one that we see spreading beyond the technology sector to politics, scientific research, education, and culture.

Diana Rawlinson

THE RADIO

3-SECOND SURVEY
Radio was the first technology to make it possible for people to participate in cultural events in real time and regardless of economic status.

3-MINUTE OVERVIEW
The growth of radio from the 1920s onward provided unprecedented access to entertainment and information and proved a powerful medium for education. Requiring no special knowledge to operate, the radio enabled people to hear about the latest societal trends. Broadcasting had an immediacy that emphasized interconnectedness and mobilized opinions. The proliferation of wireless sets in the home spawned whole new industries in program production, advertising, and newscasting.

In 1886, Heinrich Hertz showed that rapid variations of electric current could be projected into space in the form of radio waves. This breakthrough led to the first incarnation of wireless communication as Morse code-base "wireless telegraphy" patented by Guglielmo Marconi in England in 1896 and simultaneously by Nikola Tesla in the United States. In Calcutta, at the same time, J.C. Bose invented the coherer, allowing for radio waves to pass through walls or water. Due to its technical limitations, wireless transmission at this time was used principally for military and particularly maritime endeavors, where it transformed the safety of ocean travel, but had little impact on the rest of society. Additional technological improvements were needed before voice broadcasting became possible: in 1906, Reginald Fessenden broadcast a limited program of speech and music from Brant Rock, Massachusetts. Official concerns about the security of wireless transmission stifled further growth of radio during World War I, but in 1919 wartime controls were relaxed and the development of speech broadcasting began in earnest. People quickly adjusted to "tuning in" to hear about national and international events as they unfolded, ushering in the era of mass media.

RELATED TOPICS
See also
THE COMPUTER
page 98

THE TELEGRAPH
page 100

THE TELEVISION
page 106

3-SECOND BIOGRAPHIES
NIKOLA TESLA
1856–1943
Serbian-American electrical engineer who patented the alternating current induction motor; locked in dispute with Marconi over the invention of the wireless, in 1943 Tesla was posthumously awarded the patent in the United States

GUGLIELMO MARCONI
1874–1937
Italian inventor who developed the first long-distance wireless telegraph and in 1901 broadcast the first transatlantic signal

EXPERT
Diana Rawlinson

Advances in electronics in the 1900s led to wireless telegraphy becoming an entertainment industry.

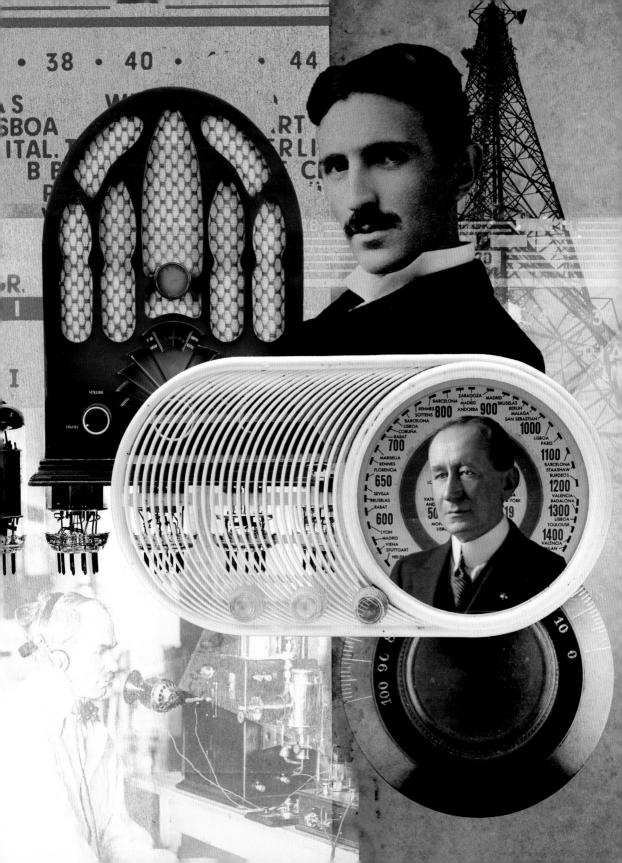

THE TELEVISION

From 1900, inventors in Britain, Japan, Germany and the United States were experimenting with ideas that would become the television, and dozens of partially successful attempts were produced. In 1925, John Logie Baird amazed London shoppers by showing the first indistinct images, but in the race to produce a commercially viable system he faced stiff competition. Within two years, Baird's "mechanical television" had been superseded by the "all-electronic" version developed by Vladimir Zworykin and Philo Farnsworth in the United States. Still, use of TV sets was not widespread, and the content limited to barely discernible figures filmed with a single camera. During World War II, developments in commercial television were abandoned or even prohibited, but after 1946 they relaunched with an improved user experience, due to technology derived from wartime innovations in radar. Ownership of sets mushroomed; money poured in from advertising and sponsorship, which became an important feature in broadcasting. Television changed what people did with their leisure time and, despite its entertainment and educational benefits, the growth of watching television attracted critics, with the recognition that whoever controlled the content wielded great power, and that its promotion of consumerism could be divisive.

RELATED TOPICS

See also
THE COMPUTER
page 98

THE RADIO
page 104

THE SMARTPHONE
page 108

3-SECOND SURVEY

Television entertains and informs us. It provides a sense of connectedness due to its "live" nature and its popularity has transformed family and social interaction.

3-MINUTE OVERVIEW

In the twenty-first century, many households include more televisions than people, and viewing has had a significant impact on how we relate and influenced the values and norms of society. Television has been blamed for everything from obesity to an increase in violence, and terms such as "boob tube" and "idiot's lantern" capture the cautionary sense that while it can be an educational medium it does not always broaden the mind.

3-SECOND BIOGRAPHIES

PAUL GOTTLIEB NIPKOW
1860–1940
German engineer who was the first to patent the "mechanical television" in 1884 using perforated rotating disks; his invention was theoretical

JOHN LOGIE BAIRD
1888–1946
Scottish inventor who demonstrated the first working television, the first color television tube, and in 1928 achieved the first transatlantic transmission

EXPERT

Diana Rawlinson

TV is arguably the most widespread invention of the twentieth century, providing a window on the world for every sector of society.

THE SMARTPHONE

The first phone to combine the functions of a cell phone with a personal digital assistant was developed by IBM in 1992 and marketed two years later by BellSouth. Named Simon, it was capable of sending and receiving emails and faxes and had features for keeping notes, storing addresses, and scheduling appointments. The addition of QWERTY keyboards made it easier to work on the move using word processing and spreadsheets. The Nokia 9000, released in 1996, included a touchscreen with stylus and offered web browsing—unlike Simon. With Penelope, the Ericsson G.S. 88 in 1997, the term "smartphone" was officially coined. The plethora of similar phones that followed shared impediments that slowed smartphone penetration of the wider consumer market: apps were limited, resistive screens required pressure that was not finger-friendly, and web pages were often truncated versions of the PC experience. In 2007, the Apple iPhone broke the mold with capacitive digitizer technology bringing a true touchscreen, and incorporating the iOS WebKit browser (quickly followed by an Android OS) allowing for full-color, completely rendered web pages. This in turn led to the development of a host of apps for everyday and social functions, transforming the smartphone from a business tool into an object of consumer desire.

3-SECOND SURVEY
The smartphone has quickly become an indispensable conduit to our digital social lives, the irony being that we have less time for face-to-face connections.

3-MINUTE OVERVIEW
Smartphones are practically ubiquitous, letting us carry out the administrative aspects of our lives and stay connected wherever we may be. This is a convenience that we have eagerly adopted, but there are drawbacks to always being reachable. It can mean that work never stops. Critics talk of overuse leading to addiction, bad posture and anxiety, and new phenomena such as trolling and cyberbullying have become the negative face of constant connectivity.

RELATED TOPICS
See also
THE COMPUTER
page 98

THE TELEVISION
page 106

3-SECOND BIOGRAPHY
THEODORE GEORGE "TED" PARASKEVAKOS
1937–
Greek-American inventor and electronic engineer who pioneered the transmission of electronic data through telephone lines; he patented the concept of combining data processing and visual-display screens with telephones in 1973

EXPERT
Diana Rawlinson

The smartphone puts the power of desktop computing into your pocket—provided you can get a signal—introducing a potential that has yet to be fully exploited.

ECONOMICS & ENERGY

American Revolution Process whereby the American colonies broke free of British rule to create the United States in 1765–83. One of the issues between the colonists and the British parliament was the use of paper money, pioneered by the founding father Benjamin Franklin.

bitcoin New electronic currency that circulates independently of central banks, and where the records of transactions are spread widely through the system. It is said to have been invented by an anonymous digital pioneer named Satoshi Nakamoto, who has never been identified. The first bitcoins were issued in 2008, created at a regular pace in return for using computing power to verify the currency, and by 2013 each bitcoin was worth more than $1,000.

blockchain Basic idea behind bitcoin and similar e-currencies, which means that the records and Internet protocols (rules for communications between devices) are distributed through the system and can't be intercepted by one user.

bubonic plague Most common form of medieval plague, probably spread by fleas on black rats, which—during the Black Death (at its peak in 1346–53)—killed one-third of Europe's population.

Chernobyl disaster Fire at the Chernobyl nuclear plant in the Ukraine in 1986 that led to the emergency resettlement of about 350,000 people and a radioactive cloud that drifted across northern Europe.

Chicago Pile-1 World's first working nuclear reactor, constructed at the University of Chicago in 1942 as part of the Manhattan Project to build an atomic bomb.

counting board Early form of abacus, using stones or counters made of wood to count on a board.

counting ropes Counting systems akin to the abacus and used by the ancient civilizations of central America, including the Aztecs. The Incas called them *quipu*.

Difference engine Name given by Victorian computer pioneer Charles Babbage to his first design, which was powered by a crankshaft and rapidly gave way to his design for an analytical engine. He never managed to build either, but the design was completed recently and is in the Science Museum in London.

digicash First digital money system, designed by David Chaum and launched in 1990. It was used later as the basis for designing secure voting systems.

efficiency Ability to use less effort, money, energy, or time to make a system work.

fossil fuels Carbon-based fuels extracted from the ground and blamed for the changing climate when they are burned.

French Revolution Uprising against aristocratic rule that burst into life in France in 1789, and blamed by some historians partly on the debasement of the French currency using paper money experiments in the early part of the eighteenth century.

Fukushima disaster Meltdown of three reactor cores that took place in 2011 at the Fukushima nuclear plant in Japan after an earthquake and tsunami. There were no immediate deaths but the plant is so toxic that it will not be safe for safety workers to decommission it for up to 40 years.

IBM Successful company that pioneered computing but missed out on the personal computer revolution.

local exchange and trading system (LETS) Barter money system in which those involved can barter—not with individuals, who might or might not have what they want—but into a central bank.

Mondex First digital "purse," capable of holding several different currencies. It was developed at the UK's National Westminster Bank, launched in 1995 and sold on later to US financial services corporation Mastercard.

nuclear reactor Power generation plant that is able to generate electricity by starting and controlling a nuclear reaction.

semiconductors Major component in most electric circuits, with conductivity less than a conductor, such as metal, but more than an insulator.

slide rule Sliding ruler used from the seventeenth century onward to carry out multiplication, division, and more complex mathematical processes.

SOS Save our souls, the Morse code distress signal, first used in 1912.

Tang dynasty Chinese monarchy from the seventh century CE to the tenth century, usually regarded as the high point of Chinese civilization.

Thirty Years' War Series of wars between Catholic and Protestant states in central and northern Europe, lasting 1618–48 and causing an estimated 8 million casualties.

THE ABACUS

3-SECOND SURVEY
Most people have ten fingers, useful for counting, but if you need to do basic calculations with more—and you don't have numerals, either—you can always use an abacus.

3-MINUTE OVERVIEW
Don't make the mistake of assuming that abacus calculations are slow. During the military occupation of Japan by the United States from 1945, American officials were fascinated by the speed with which people used the abacus, so they carried out an experiment. Their electronic calculator operators were beaten by locals using the abacus for all mathematical operations apart from division.

Nearly every ancient culture used something like an abacus to do mathematical calculations—right back to the old Babylonian empire, which used a counting system based on the number 60. No abacuses or counting boards from that era (c. 1900 BCE) have been found, probably because they were made of wood and have rotted away, but images of them do survive. Later on the ancient Egyptians, Greeks, Chinese, and Romans all used a device like an abacus, moving beads quickly across a board to add or subtract, while the Aztecs in Central America used a similar gadget with knots on a piece of cord to help them with their number system, which was based on 20. There is some evidence that the Roman abacus was copied from the Chinese one; not only were the calculating machines very similar, but there were also trading links between the two countries. The problem in Europe was that Roman numerals were difficult to calculate with. The mathematician Gerbert of Aurillac, who became Pope Sylvester II, reintroduced the abacus into Western Europe, having learned about Arabic numerals from Arab scholars in Cordoba and Seville in Spain—though he did so without a numeral for zero. The abacus lasted until the age of the slide rule and pocket calculator.

RELATED TOPIC
See also
PAPER MONEY
page 116

3-SECOND BIOGRAPHIES
GERBERT OF AURILLAC
(POPE SYLVESTER II)
946–1003
French mathematician who introduced the Arabic numeral system to Europe and brought with it the abacus

TERENCE V. ("TIM") CRANMER
1925–2001
American founder of the International Braille Research Center and inventor of the Cranmer Abacus, used to this day by blind people

EXPERT
David Boyle

Fast and effective, the abacus led to the development of Arabic numerals.

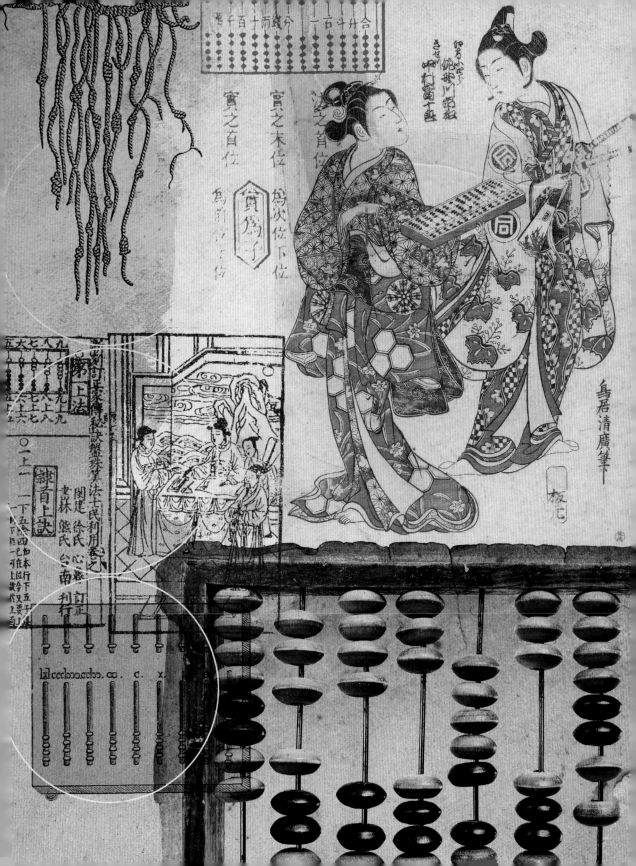

PAPER MONEY

Imagine you had spent your whole life, as your ancestors had spent theirs, regarding money as an item that had a value in its own right. Then suddenly, you were offered a piece of paper representing money instead. That was what happened to Venetian explorer Marco Polo when he reached the court of Kublai Khan at some point in the 1260s. He was the first European to encounter the phenomenon of paper money, noting with astonishment that it wasn't an option to refuse it. The Great Khan insisted and his insistence carried the force of law. Nobody knows who actually invented paper money. It seems to have evolved from receipts and promissory notes, emerging for the first time during the Chinese Tang dynasty in the seventh century CE, and again when copper was in short supply for coins—which in any case Chinese merchants found too heavy to carry around. The first European banknotes emerged as late as 1661 in Stockholm, as a promise to pay in gold or silver. Yet delinking money from an intrinsic value did carry a huge risk; the inflation and sudden deflations could be terrifying. In fact, a Dutch merchant, Johan Palmstruch, who introduced those first notes in Sweden, was condemned to death for irresponsible book-keeping and died in prison.

3-SECOND SURVEY

Paper money boldly delinks money from the intrinsic value it represents, making it cheap to produce, usually out of paper or plastic.

3-MINUTE OVERVIEW

Paper money separates money from gold, which is generally regarded as a good thing, but it can also—as the economist John Maynard Keynes warned—be a weapon of mass destruction. Both the American and French revolutions have been blamed on paper money. In fact, the French aristocracy was turned completely upside down by the boom and bust ushered in by the Scottish economist John Law's issue of banknotes in France in the early eighteenth century.

RELATED TOPICS

See also
E-MONEY
page 128

THE CREDIT CARD
page 130

3-SECOND BIOGRAPHIES

KUBLAI KHAN
1215–94
Ruler of the Mongol empire, one of the first recorded rulers to have specifically ordered paper money

JOHN LAW
1671–1729
Scottish gambler and paper money pioneer who came close to bankrupting France with his experiments

BENJAMIN FRANKLIN
1706–90
American founding father who made the case for paper money to a suspicious British House of Commons

EXPERT

David Boyle

China had used paper currency for five centuries before European countries followed suit.

NUCLEAR POWER

RELATED TOPIC
See also
THE WIND TURBINE
page 124

3-SECOND SURVEY
If splitting the atom could produce energy, it made sense to harness it for the good of the world—but it hasn't always proved easy.

3-MINUTE OVERVIEW
Two nuclear accidents have thrown the development of civil nuclear energy into doubt—Chernobyl in the Ukraine in 1986, where safety systems were shut down, and Fukushima in Japan in 2011, the result of an earthquake and tsunami. In both cases, immediate deaths were relatively low but the land around will not be usable for two generations and the costs of cleanup and precautions at other plants may be immeasurable.

President Dwight Eisenhower's administration first used the phrase "atoms for peace" as part of an attempt to make nuclear weapons seem a little more hopeful. It was a reaction against the helter-skelter rush to develop nuclear technology during World War II, to beat the Nazis to the key. It had been an eventful story since the physicist Ernest Rutherford first bombarded lithium atoms with protons and found that a huge amount of energy was released. The critical experiments were carried out in Germany in 1938 by Otto Hahn and Fritz Strassmann, with Austrian physicist Lise Meitner and her nephew, Otto Frisch. But within a year, the war had begun and many of the world's best physicists were refugees in the United Kingdom or the United States. The first working reactor went live as an experiment in 1942 and was called Chicago Pile-1; this fed into the Manhattan Project, designed to make a nuclear bomb. The first working nuclear reactor to actually produce any electricity was the small Experimental Breeder Reactor in Idaho in 1951. Since then, nuclear energy has struggled to live up to early expectations, mainly because of the sheer expense.

3-SECOND BIOGRAPHIES
ERNEST RUTHERFORD
1871–1937
New Zealand-born physicist whose work led to the first splitting of the atom

OTTO HAHN
1879–1968
German nuclear physics pioneer, who contemplated suicide when he discovered that atomic bombs had been dropped on Hiroshima and Nagasaki

FRITZ STRASSMANN
1902–80
German chemist who helped Otto Hahn discover the phenomenon of nuclear fission

EXPERT
David Boyle

In spite of being heralded as "clean energy," accidents and expense have given nuclear power an uncertain future.

1883
American inventor
Charles Fritts creates
a solar cell using gold
and selenium; it is around
1 percent efficient

January 31, 1898
Russell Ohl born near
Allentown, Pennsylvania

February 23, 1940
Works on the silicon
crystal with a crack down
it that leads to his
surprise discovery

1946
Gets his patent for solar
cells

January 23, 1948
Discovers a transistor
based on Ohl's N-P
barrier

April 25, 1954
Bell Laboratories
announce the invention
of the first practical
silicon solar cell

May 14, 1973
Launch of the space
station Skylab, powered
by solar panels

March 20, 1987
Dies in California

2008
A new record is broken
in solar cell efficiency
when the US National
Renewable Energy
Laboratory creates a
cell generating power
at more than 40 percent
efficiency

RUSSELL OHL

During his first year at college,
Russell Ohl heard his very first radio message.
It was an SOS from a ship out in the Atlantic
being attacked by a U-boat during World War I.
He had gone to Pennsylvania State University at
the age of just 16, and enrolled on the fearsome
electrochemical engineering course—at the end
of the first year, only three of the original
students remained. The radio transmission had
caught his imagination and he was determined
to go into radio technology.

As a result of that fascination, Ohl stumbled
on a peculiarity in silicon that made the element
the basis for the invention of transistors and
also, more directly, of the modern world of solar
cells—getting sunlight to generate electricity.

Ohl began work in the Army Signal Corps
and progressed through laboratories at
Westinghouse and AT&T before he finally began
at Bell Laboratories in New Jersey. It was a
frustrating career. He was fascinated by the
possibilities of semiconductors that control
electrical currents, but was constantly being
moved from that work to something more
urgent, because it was an Air Force contract or
something equally vital.

"There were a great many enemies to this
work with semiconductors," he said later. "You
have no idea how many people opposed that.
Vacuum tube people said that there is nothing
to it and it is all a lot of tommyrot, and that sort
of thing."

The problem was that existing radio tuners
didn't work well at high frequencies, so he
began to experiment with semiconductor
materials, such as silicon.

In 1940, while working on a silicon sample
with a crack down the middle, Ohl found to his
surprise that the current flowed between the
two sides when it was exposed to light. By a
peculiar error, the silicon atoms on one side
seemed to have more electrons than those
on the other. The gap became known as the
"N-P barrier" (negative-positive). He found
he had made a simple device that turned
sunlight into electricity.

While there had been solar cells before using
selenium, Ohl's cell was far more efficient.
There are those who forecast that solar cells
will power the next economic revolution.

David Boyle

THE CALCULATOR

Electronic pocket calculators were the must-have tools of every executive in the 1970s. The invention transformed daily lives by greatly expanding the math capabilities of everyone from students to business people. It had taken 2,000 years to progress from the abacus, the ancient world's hand-operated "calculating board" using beads on rods, to mechanical calculators. Wilhelm Schickard invented the "calculating clock" in 1623, a mechanical adding/subtracting calculator operated by a series of gears. A device invented by the philosopher Blaise Pascal in 1642 also used geared wheels and could add and subtract two numbers directly and multiply and divide by repetition. In the following century, Charles Babbage produced what he called a difference engine, forerunner of the computer. Mechanical calculators, such as the first commercially viable counting machine—the Arithmometer (in production until 1915)—and the US-patented push-button Comptometer, dominated office life up until the 1960s. Yet in barely a decade the calculator was transformed; a bulky, heavy desktop machine that ran on household electricity, and cost more than a family car, became a cheap and compact battery-power device that would slip into a pocket. Innovations such as integrated microchips and electronics that could run on batteries made this transition possible.

3-SECOND SURVEY
Abacus, slide rule, and then calculating machine—the world seems to have beaten a path toward the pocket calculator, until it was driven out by computers.

3-MINUTE OVERVIEW
When Clive Sinclair introduced his Sinclair Executive in 1972, it was the first truly pocket-size calculator. They sold for $200 each. By the end of the decade, similar machines cost just around $11 and the calculator industry was about to be blown away by the development of personal computers.

RELATED TOPICS
See also
THE COMPUTER
page 98

THE ABACUS
page 116

3-SECOND BIOGRAPHIES
WILHELM SCHICKARD
1592–1635
German-Hebrew professor and astronomer credited as creator of the first calculator

BLAISE PASCAL
1623–62
French philosopher who developed the first working calculating machine

CLIVE SINCLAIR
1940–
British inventor and pioneer of the first pocket calculator

EXPERT
David Boyle

The pocket calculator evolved from the abacus over four thousand years, but technological advance is hastening its demise.

THE WIND TURBINE

Windmills were the pioneering technology of the twelfth century in Europe, harnessing the power of the wind to mechanically turn grinding machines to make flour and drive small machines. The idea that they might also generate energy had to wait, needless to say, until the invention of electricity. After that, it was the Scots who pioneered the idea, perhaps not surprisingly given the prodigious amount of wind that blows across Scotland. It was a Scottish teacher, James Blyth, who set up the first battery-charging wind turbine to power his vacation home in Maykirk, Kincardineshire, in 1887. Wind generators made of steel were a common sight on farms and remote places not on the grid in the 1930s, particularly in the United States. There were also efforts to plug wind energy into the grid in places such as Yalta in the 1930s, Vermont in the 1940s, and the Orkneys in the 1950s. But fossil fuels continued to dominate until demand for renewable energy sources put wind energy back on the map, starting in the 1970s. Most modern commercial wind turbines are based on a Danish model, with three lightweight horizontal blades that can swivel to meet the wind at the best angle for harvesting this energy. A generator inside the main turbine body takes kinetic energy from the spinning drive shaft (to which the blades are attached) and turns it into electrical energy.

RELATED TOPIC
See also
NUCLEAR POWER
page 118

3-SECOND SURVEY
In windy places, it makes sense to find a technology to make sure that the energy in the wind is put to good use.

3-MINUTE OVERVIEW
Wind power cannot be enough on its own because many places have unreliable wind. In some places it blows too fiercely and the windmills may topple. Sometimes the wind doesn't blow at all. So wind power, important as it is—and it can boost the energy independence of communities and nations—does need to be balanced by other sources of energy, too.

3-SECOND BIOGRAPHIES
JAMES BLYTH
1839–1906
Scottish math teacher who first generated energy from the wind

PALMER COSSLETT PUTNAM
1900–84
American geologist responsible for the first American wind turbine, built in Vermont in 1941, which broke down after 1,000 hours

EXPERT
David Boyle

Offshore wind farms, most recently built on floating rigs, harness higher offshore wind speeds and attract less opposition than onshore construction.

THE BARCODE

3-SECOND SURVEY

The barcode—a simple pattern of black lines of varying thicknesses that can be scanned electronically—provides the rapid solution to identifying one item among millions.

3-MINUTE OVERVIEW

Woodland and Silver were not the only people working on the idea. Unknown to them, an executive from Pennsylvania Railroad was trying to work out how you could identify carriages or trucks instantly. David Collins tried a system of blue and red lights called KarTrak. The idea did not take off there but it was adopted instead by a New Jersey toll road that was trying to work out which cars had paid their subscriptions.

A conversation overheard between a supermarket executive and a senior lecturer at the Drexel Institute of Technology in Philadelphia in 1948 led student Bernard Silver to pursue the idea of barcodes. The question was about how to develop a system that could identify a product automatically. His first attempt at an answer—with his friend and fellow-student Norman Joseph Woodland—involved ultraviolet ink, which turned out to be too expensive. The two friends had their barcode system patented in 1952, by which time Woodland was working to interest IBM in the idea. Ten years later, they sold the rights for just $15,000. It was only when the American electronics giant RCA bought the patent and began to develop a similar system that IBM started to catch up. The first grocery item to be scanned was a bar of chewing gum in Troy, Ohio, in 1974. The gum and the receipt are now on display in the Smithsonian Institution in Washington, DC, and barcodes are, of course, absolutely ubiquitous—on books, cars, letters, package, and passports. They have become a lampooned and central part of modern life. Neither of the inventors made any more money out of it. Silver died of leukemia in 1963 and Woodland watched bemused as their creation took off across the world.

RELATED TOPIC
See also
THE COMPUTER
page 126

3-SECOND BIOGRAPHIES
NORMAN JOSEPH WOODLAND
1921–2012
American lecturer in mechanical engineering who first developed barcodes by drawing a series of lines in the sand

BERNARD SILVER
1924–63
American electrical engineer who came up with the original idea for the barcode

EXPERT
David Boyle

The simple barcode helps us keep track of everything from travel tickets to shopping.

E-MONEY

When banks started creating money in the form of loans with a stroke of a pen in a balance sheet—many centuries before the advent of information technology—this went some way to making money the virtual idea it is today. But in those days cash was still represented by a thing—a note or a coin. A cryptologist from Los Angeles, David Chaum, is credited with having invented the first digital money, which he imagined in 1981 and then developed and launched under the title DigiCash. It had built-in coding to protect the money and to make it untraceable—most e-money is not these days. Chaum moved on when Digicash unraveled in 1998, and used his ideas instead for voting systems, where similar problems of protecting personal identity needed to be solved. He was followed close behind by two innovators from the National Westminster Bank in London, Tim Jones and Graham Higgins, who invented the first electronic purse, using (among others) a digital money called Mondex. Mondex was sold to Mastercard but it failed to spread as rapidly as expected. As the digital money pioneers put it: There was a need for their inventions, but not nearly enough demand.

RELATED TOPICS
See also
PAPER MONEY
page 116

THE CREDIT CARD
page 130

3-SECOND SURVEY
Cash can be heavy to carry, dirty to count, and expensive to guard; it may be more convenient to use blips on a screen.

3-MINUTE OVERVIEW
E-money allows for you to exchange small sums with a click. Pioneers include Michael Linton in Canada, with his Local Exchange and Trading Systems (1983), and the mysterious and unidentified Satoshi Nakamoto, who launched bitcoin (2008), using the so-called "blockchain" technology that encrypts the ledger in every transaction.

3-SECOND BIOGRAPHY
DAVID CHAUM
1955–
American cryptologist and mathematician who launched the first digital money, DigiCash

EXPERT
David Boyle

Ethereal perhaps but e-money needs to be as secure as a bank vault.

THE CREDIT CARD

RELATED TOPIC
See also
E-MONEY
page 128

It all began with a lost checkbook.

Financier Frank McNamara was having lunch with his friend Alfred Bloomingdale at Major's Cabin Grill restaurant, next to the Empire State Building in New York City, when they found neither of them had enough cash to pay the bill. The story goes that McNamara had to phone his wife and ask her to drive around with some money. But it was also a revolutionary moment. The two men spent the rest of the meal discussing an idea that could prevent that kind of inconvenience arising again. The following year, in 1950, McNamara launched the Diners' Club, the world's first credit card. But the real shift happened eight years later when a Bank of America middle manager, Joe Williams, organized the famous Fresno Drop, which involved sending 60,000 credit cards out to the inhabitants of Fresno, California, whether they had asked for one or not. It wasn't just the beginning of credit cards in the mass market—it was also the beginning of easy credit for the masses in the United States, which jump-started the consumer revolution. The Diners' Club was an immediate success but McNamara sold his interest a couple of years later and his business is said to have lost a great deal of money in the property market.

3-SECOND SURVEY
Credit makes money a flexible thing; credit cards made what was once just for the wealthy available to almost everyone—at a price.

3-MINUTE OVERVIEW
If McNamara produced the first credit card, the idea emerged much earlier in an 1888 futuristic novel by American writer Edward Bellamy, called *Looking Backward 2000-1887*. He imagined that everyone would get a money card on which the government would load their monthly allowance. The book was highly controversial and so horrified the British artist William Morris that he wrote his own version (*News from Nowhere*) without credit cards.

3-SECOND BIOGRAPHIES
EDWARD BELLAMY
1850–98
American writer who first imagined credit cards in his novel about the year 2000

FRANK MCNAMARA
1917–57
American financier who launched Diners' Club and the whole idea of credit cards

EXPERT
David Boyle

The idea for the credit card began in a restaurant and now almost everyone in the developed world uses one.

DAILY LIFE

amphorae Large, mostly earthenware, handled containers used for the storage and transport of drink and foodstuff, typically wine and olive oil. Used for thousands of years from the Neolithic period onward, they had an airtight seal for preservation; often they had pointed bases, which meant they could be stood upright in soft ground or stored several layers deep aboard trading ships.

arc lamps Extremely bright lights first made in the early 1800s by passing an electric current in air between two electrodes made from carbon. Widely used for lighting streets and factory buildings until replaced by incandescent lights, they remained in use longer for applications such as cinema projection.

bitumen Sticky, black, viscous liquid that occurs naturally as a mix of hydrocarbon substances and is manufactured artificially as a by-product of petroleum distillation.

camera obscura Device used for projecting an external view into a closed, darkened space. A camera obscura can range in size from a room to a small box—and works by letting light pass through a small hole, in the process inverting the image.

Cinématographe Early example of a convergence device that performed several different functions and was the first mechanism to provide a modern movie theater experience in which multiple people could simultaneously watch projected, moving images. It combined camera, printer, and projector in one unit.

cooling by vapor compression Process of cooling used in refrigerators and air-conditioning systems. It uses a contained, liquid refrigerant to absorb and remove heat from a given space through a cycle of vaporization, compression, and expansion. The same process works whether inside a refrigerator or a room in a building.

enzymes Biological catalysts that speed up chemical reactions. They each operate best at a particular temperature and in an environment with a particular acidity, or pH level, and are important in the process of respiration. Enzymes are proteins in nature and accelerate reactions by offering a pathway that requires less energy for the reaction to work.

fluorescent bulbs Bulbs that work by passing electric current through mercury vapor (a toxic substance), creating shortwave ultraviolet light that in turn causes a phosphor coating inside the bulb to glow. They make light in a more energy-efficient manner than incandescent bulbs.

Kinetograph First commercial movie camera—a mechanism for synchronizing the uniform movement and exposure of photographic film invented by William Kennedy Dickson.

Kinetoscope Solo mechanism for the projection and viewing of film taken with the Kinetograph.

LED bulbs Bulbs in which electricity is passed through a semiconductor that then emits visible light using a property called electroluminescence—whereby phosphorescent substances emit light particles in response to weak electrical current. LED bulbs produce light more efficiently again than both fluorescent and incandescent lightbulbs.

persistence of vision Ability of the brain to retain an image for slightly longer than the time for which it is seen.

phi phenomenon Ability of the brain to edit together a sequence of still images and create the illusion of continuous movement. It does this by filling in the gaps of information missing between them.

Voltaic pile First electric battery. It was made by piling copper and zinc disks, one on top of the other, separated by cloth soaked in a salt or "saline" solution. This worked as an electrolyte—a substance that in solution can conduct electricity—created by differences between the metals.

zero-carbon houses Dwellings that in their construction and operation result in no net carbon emissions being released into the atmosphere. This means, in reality, that their materials must be very low-carbon and they must be extremely energy-efficient—utilizing passive solar energy and other renewable sources. To be truly zero carbon, however, some method for storing or "sequestering" carbon will probably be needed.

zoetrope A spinning cylinder with uniform viewing slots cut into it that contain static images, which when viewed through the slots, as the cylinder spins, create the illusion of moving images.

THE LAVATORY

Globally, 2.4 billion people lack access to sanitation, according to the World Health Organization. Of those, more than 940 million defecate in the open. It's a shocking statistic for everyone who takes having a toilet for granted. Domestic toilets connected to some form of drainage or cesspit are almost 5,000 years old. Public bathrooms integrated into urban sewers were important meeting places in ancient Rome. But the flushing toilet is a more recent invention. The first working version is thought to have been designed by English courtier Sir John Harington in 1596 for his godmother Queen Elizabeth I. But the flushing mechanism at the heart of the invention was invented much earlier as part of a hand-washing automaton by the Muslim polymath and mechanical engineer Ismail al-Jazari, who described his invention in 1206. Flushing toilets wouldn't reach a broad population, however, until the Industrial Revolution. A patent for one incorporating the crucial Sbend was filed by Alexander Cumming in 1775. Another inventor and contemporary of Cumming's, Joseph Bramah, oversaw improvements to the design, which had a tendency to freeze in cold weather. History best remembers, however, the man who in the nineteenth century improved the flushing toilet's mechanism and popularized it: Thomas Crapper.

3-SECOND SURVEY

Two-thirds of the world's population can flush away their waste problem, but for the rest the door to a solution remains locked and "engaged."

3-MINUTE OVERVIEW

The United Nations' Sustainable Development Goal Number Six is to "ensure access to water and sanitation for all." Achieving universal coverage has been costed at just over $500 billion, or less than the annual amount spent on its military by the United States. Some charities, such as the Bill & Melinda Gates Foundation, invest in high-tech answers to the world's toilet problem; for others, the most sustainable solution is the humble compost toilet.

RELATED TOPIC

See also
ANTIBIOTICS
page 76

3-SECOND BIOGRAPHIES

ISMAIL AL-JAZARI
1136–1206
Turkish polymath, inventor of a handwashing automaton

JOHN HARINGTON
1561–1612
English courtier, author, and inventor of an early flushing toilet

EXPERT

Andrew Simms

In 1596, Queen Elizabeth I became the first monarch to "sit on the throne" in the modern sense.

CANNING

A 3,000-year-old oak barrel of butter was dug up in 2009 in County Kildare, Ireland. It looked uncannily like a rusty food can. Packaging food for preservation is something people have done for millennia. The Romans preserved fish pickle sauce, olive oil, wine, and other foods in sealed amphorae, pipes and wells. Heating food before placing it in airtight containers preserves it by killing microorganisms and "switching off" enzymes that would lead to its decay, creating a new protective shield. But it wasn't until the 1860s, and Louis Pasteur's work on microorganisms, that sterilization was understood. The basic practice of this method of preservation—so-called "thermal processing"—was discovered by a French confectioner without realizing the science at work. In 1795, Nicolas Appert began experimenting with airtight containers using glass bottles and his discovery was later employed by the French navy, which needed a method of preserving food for long voyages. English inventor Peter Durand used a mixture of ceramic and glass in his storage vessels and, in 1810, patented food containers made from iron coated with tin. Food in tin cans was made available in the United States in 1822 by Thomas Kensett and Ezra Daggett. Today cans are mostly made from steel, with around 100 million used daily in the United States alone.

3-SECOND SURVEY

Tin cans revolutionized the storage and preservation of food in the nineteenth century and have been popular ever since.

3-MINUTE OVERVIEW

Recycling one metal can saves the equivalent amount of energy to that used by a television in three hours, or 74 percent of the energy needed to make the can in the first place. Globally, only about 8 percent of tin actually gets recycled. From China to Indonesia, tin mining and smelting is heavily polluting the environment. Lighter and more environmentally friendly forms of food packaging now increasingly replace metal cans.

RELATED TOPICS

See also
STEEL
page 24

LOUIS PASTEUR
page 80

REFRIGERATION
page 148

3-SECOND BIOGRAPHIES

NICOLAS APPERT
1750–1841
French confectioner and distiller who experimented with preserving food in heated, sealed bottles

PETER DURAND
1766–1822
English merchant and inventor who was given a royal patent for preserving food in tin-lined iron cans in 1810

EXPERT

Andrew Simms

Canning may seem like a modern solution to extending the life of perishable food but the invention has been around for much longer than many realize.

THE LIGHTBULB

3-SECOND SURVEY
Electric lighting transformed the world, irrevocably changing the ease with which people could work, socialize and travel after dark.

3-MINUTE OVERVIEW
The Livermore Centennial Bulb is thought to be the world's oldest, having glowed almost continuously since 1901. But it is an exception. In the 1920s, a manufacturers' cartel known as Phoebus, which included Osram, Philips and subsidiaries of General Electric, conspired to produce bulbs with shorter lives than was then common—just 1,000 hours on average instead of 1,500–2,000. Today's energy-efficient compact fluorescent bulbs can last 8,000 hours, and LED bulbs up to 50,000.

Ironically the invention of the incandescent lightbulb was more of a slow dawn than a "lightbulb moment." While Thomas Edison receives most of the credit for key lightbulb patents filed by researchers in his company, these stand in the afterglow of a number of earlier inventions. At the turn of the nineteenth century, Italian physicist and chemist Alessandro Volta developed a zinc and silver battery—or "voltaic pile"—with a glowing copper connecting wire. His invention prefigured the bright filaments of Edison's bulbs by 80 years. Extremely bright "arc lamps"—connecting Volta's batteries with charcoal electrodes—were developed by English inventor Humphry Davy in 1802. But it took many more decades and advances in engineering before a viable household bulb was produced. By 1840 a costly bulb with platinum filament was available. But it required the right filament material and the creation of a vacuum to make an efficient bulb that lasted. After decades of experimentation, Edison bought patents from others trying to commercialize lightbulbs and tried filaments made from wood, cotton, and thousands of plant materials. His team filed a key patent in 1879 and discovered that carbonized bamboo filaments lasted for more than 1,000 hours. Tungsten filaments, introduced around 1910, were used until incandescent bulbs fell out of fashion.

RELATED TOPICS
See also
THE CAMERA
page 142

MOVING PICTURES
page 146

3-SECOND BIOGRAPHIES
ALESSANDRO VOLTA
1745–1827
Italian physicist, chemist, electrical pioneer, and inventor, who developed the battery

THOMAS ALVA EDISON
1847–1931
American inventor and businessman, who patented the first commercially successful lightbulb

THE ELECTRICAL ASSOCIATION FOR WOMEN
1925–86
UK organization formed by the Women's Engineering Society for "popularizing the domestic use of electricity" and to promote "electrical housecraft"

EXPERT
Andrew Simms

Economic self-interest held back the full benefits of early household lighting.

THE CAMERA

The basic mechanics of the camera are older than most suppose. The way that light passing through a small hole into a darkened space can project an image of what is outside was known to the ancient Greeks and noted by Aristotle. It is the discovery of the chemical process to fix the image that comes much later. The *camera obscura* (Latin for "darkened room"), ranged in size from a whole room to a box. These cameras were used as an entertaining novelty and a drawing aid for artists for centuries before the arrival of modern photography. That began with the experiments in heliography, or "sun drawing," by the Frenchman Joseph-Nicéphore Niépce. After years of chemical experimentation, in 1826–27, he combined a camera obscura and a metal plate covered with a kind of light-sensitive bitumen to make what is considered the first permanent photograph from life—a view from his workshop. The process was slow and difficult. Photography's subsequent development involved reducing the size of the equipment and enhancing the speed and quality of image reproduction. In 1861, a collaboration between English photography pioneer Thomas Sutton and Scottish physicist James Clerk Maxwell produced the first color photograph. Three negatives separated red, green, and blue light in a way that still underpins chemical and electronic methods for recording images.

3-SECOND SURVEY
For millennia people knew that light passing through a small hole to a dark space projected an image; chemistry then allowed for it to be captured.

3-MINUTE OVERVIEW
In its early days, photography could dispel myths about everything— from imagined foreign lands to the reality of war. But it could also create new myths, with its apparently unquestionable accuracy. The camera became a source of wonder in the hands of travelers and a tool of control in the hands of the powerful. Now, with a camera in every smartphone, almost everyone is a photographer.

RELATED TOPICS
See also
THE TELEVISION
page 106

THE SMARTPHONE
page 108

MOVING PICTURES
page 146

3-SECOND BIOGRAPHIES
JOSEPH-NICÉPHORE NIÉPCE
1765–1833
French inventor who devised the first method for recording a permanent photographic image

THOMAS SUTTON
1819–75
English inventor who took the first color photo in 1861, using Maxwell's method

JAMES CLERK MAXWELL
1831–79
Scottish physicist whose work on human perception of color led to color photography

EXPERT
Andrew Simms

A period of 250 years took us from chemistry in a "darkened room" to "selfie" overexposure.

1897
Father is murdered weeks before her birth

January 10, 1898
Born in Schenectady, New York

1901
Family moves to France

1912
Family moves back to New York; she attends the Rayson School and becomes interested in mathematics

1917
Awarded a degree in physics from Bryn Mawr College

1918
Completes a masters degree at the University of Chicago and becomes the first woman scientist employed at General Electric's research laboratories

1919
Publishes her first scientific paper, at the age of 21, on the use of charcoal in gas masks to filter toxins

1926
First woman awarded a PhD in physics by Cambridge University, studying in Ernest Rutherford's Cavendish laboratory

1933
Develops a "color gauge" allowing the measurement of layers of film a single molecule thick

1935
Develops a method for spreading layers of film one molecule in thickness

1938
Devises the coating method to produce nonreflective "invisible glass"

1939
Her invention is used in the filming of the movie *Gone With the Wind*, which is lauded for the quality of its images

1951
After multiple honorary degrees and other awards, becomes the first industrial scientist to be awarded the Francis Garvan Medal from the American Chemical Society

1953
Omitted from an article in the journal *Science*, celebrating 75 years of research at General Electric's laboratories

1963
Retires from General Electric

October 12, 1979
Dies at the age of 81 in Schenectady, having been active in the Zonta Club whose mission was to support and advance the status of professional women

KATHARINE BURR BLODGETT

The legacy of the inventor

Katharine Burr Blodgett is all around you. You probably spend hours every day staring at it, without even noticing. You see it, or rather you don't, whenever you look through certain windows, camera lenses, or windshields; at shop displays or cell phones; at pictures hung on walls in an art gallery; and in many other circumstances. Blodgett developed a process to eliminate the distorting glare of the reflections from glass; modern society sees more clearly thanks to what was dubbed her "invisible glass."

Born in 1898 in Schenectady, New York, Blodgett never met her father, who had been killed by a burglar shortly before she was born. But his influence lived on—she ended up working for the same company as he had done: General Electric. She was encouraged to take up scientific research by one of her father's company acquaintances, Irving Langmuir, who would later win the Nobel Prize in chemistry (in 1932) and whose assistant she would become after graduating from the University of Chicago in 1918.

Later, in 1924, she returned to studying, at Cambridge University in England, where she worked with Ernest Rutherford and became the first woman to be awarded a PhD in physics by the university. Her earlier work with Langmuir investigated the making of ultrathin, uniform oily layers on water; subsequently she worked on how to transfer the oily films onto hard surfaces and build up multiple layers, which were eventually called "Langmuir-Blodgett films." It was only after she developed the process that a practical use for it emerged. Building up a precisely calibrated number of layers of the film on glass, she discovered, had the effect of radically reducing reflection and glare. Invisible glass wasn't her only contribution; during World War II, she worked on protective measures against poison gas, the deicing of aircraft wings and smokescreens for ground troops.

Beyond research she engaged in theater, astronomy, and gardening and had two long-term relationships, at different times, with Gertrude Brown and Elsie Errington. Blodgett was given many awards and honorary degrees in her life, but went unmentioned in a 75th anniversary celebration of General Electric's research department published in the journal *Science* in 1953.

Andrew Simms

MOVING PICTURES

Just as the *camera obscura* came
before the modern camera, another entertainment
device, the zoetrope—a spinning drum with gaps
cut in its side—prefigured moving pictures. It's a
clue to the fact that the real, less acknowledged
inventor of moving pictures is the human brain.
The zoetrope allowed for quick sight of
successive still images that the brain, using
persistence of vision (in which an image is
retained slightly longer than it is seen) and the phi
phenomenon (by which a series of still images is
seen as continuous motion) combined to create
the illusion of movement. A horse galloping was
the first moving photographic picture, produced
in 1877 when Eadweard Muybridge, fascinated by
the study of animal movement, took a sequence
of pictures of a horse. He mounted them in a
zoetrope-type spinning disk: when viewed, these
settled a long-held debate, proving that all four
of a galloping horse's hooves leave the ground
simultaneously. The invention in 1888 of celluloid
film and the Kinetograph, a mechanism for
synchronizing the movement and exposure of
film commissioned by Thomas Edison from
technician William Kennedy Laurie Dickson,
were the next steps forward but they only
allowed for solo viewing on a Kinetoscope.
The technology for projection—known as the
cinématographe—was invented by Auguste
and Louis Lumière and launched in 1895.

3-SECOND BIOGRAPHIES
EADWEARD MUYBRIDGE
1830–1904
English photographer whose
studies of animal motion
created the first photographic
moving images

ALICE GUY-BLACHÉ
1873–1968
Early French film pioneer
who experimented with sound,
color, special effects and
multiracial casting, and maker
of possibly the first narrative
film, *The Cabbage Fairy*,
in 1896

EXPERT
Andrew Simms

*When reality is
either too much or
not enough, moving
pictures take us
somewhere else.*

REFRIGERATION

Heat accelerates decay, therefore finding ways to keep things cold has been a priority in human history. Before modern refrigeration, there were old-fashioned ice houses; Arctic Inuit people had little problem preserving food for long periods using ice. Domestic refrigerators are a twentieth-century invention, but artificial cooling using a process called "vapor compression" to remove heat from a confined space can be traced to demonstrations by Scottish physician William Cullen in the 1740s. It was a century later that commercial cooling systems were being introduced for industry, for example, by journalist and mechanical engineer James Harrison, who developed refrigeration for the meat industry, which was patented in 1856 and in use soon after. Today's domestic refrigerators meanwhile emerged in the nation that is still today synonymous with household appliances: the United States. Fred W. Wolf, son of an industrial refrigeration pioneer in Fort Wayne, Indiana, introduced a commercial electric household refrigerator in 1913. A year later in Detroit, Michigan, Nathaniel B. Wales developed a unit he named the "Kelvinator" after physicist William Thomson (Lord Kelvin), who discovered absolute zero. With many similar products coming onto the market by then, it is impossible to say that the refrigerator had a single inventor.

3-SECOND SURVEY

A Scottish physician's curiosity about the ability of gases and partial vacuums to remove heat led toward refrigeration.

3-MINUTE OVERVIEW

Artificial refrigeration makes life much more comfortable, but comes at a huge environmental cost. Impacts range from waste built up by refrigerator disposal to the amount of energy demanded and the heavily polluting chemicals used. When it was discovered that the chemicals used in modern refrigeration attacked our protective ozone layer, new ones were found. But these turned out to cause global warming. Cool ways to keep cool are still being sought.

RELATED TOPIC
See also
CANNING
page 138

3-SECOND BIOGRAPHIES
WILLIAM CULLEN
1710–90
Scottish physician and celebrated teacher, author of the essay "Of the Cold Produced by Evaporating Fluids and of Some Other Means of Producing Cold"

FRED W. WOLF, JR.
1879–1954
American inventor, first to develop and mass market a household refrigerator, known as the DOMELRE, short for Domestic Electric Refrigerator

EXPERT
Andrew Simms

Refrigeration made food last longer, but wasn't an environmental free lunch.

CENTRAL HEATING

One reason for human dominance on Earth is our ability to survive in a wide range of environments, from hunting on permafrost to trekking in deserts. But to really thrive, we had to learn to control the temperature of our homes. Before advanced engineering, this meant smart building to suit the environment—thick walls for insulation and holes in the right places for ventilation. Then came central heating; it's older than you think. Romans used underfloor and wall heating, blasting hot air from furnaces through clay pipes, a trick some ancient Greeks used, too. However, demonstrating that progress is never a straight line, the idea of central heating was largely lost for centuries. Leap forward to the heart of the Industrial Revolution in Derby, England, for its recognizably modern reinvention. In 1793, mill owner William Strutt developed a hot-air furnace to heat his factory, which was then adapted to improve conditions in hospitals. The Scottish inventor and steam pioneer James Watt developed a working household system of steam heating that was also used in greenhouses. But the award for most iconic invention of central heating—the hot-water radiator—goes to Franz San Galli, a Russian inventor and businessman. His 1857 radiator patent fostered our enduring modern expectation of constant, centrally heated comfort.

3-SECOND SURVEY
For millennia, civilization's elite knew how to keep warm, but modern central heating got us all hooked on constant comfort.

3-MINUTE OVERVIEW
The comfort of central heating comes at a cost. Like the energy for the Industrial Revolution, all the warm air, steam and hot water used to control the temperature of our homes has been produced by burning fossil fuels. Keeping our homes warm and at constant temperatures has ironically fueled the rapid warming and destabilization of our climate. Now the future of home comfort lies with "zero-carbon houses" warmed by renewable and passive solar energy.

RELATED TOPICS
See also
THE STEAM ENGINE
page 46

THE INTERNAL COMBUSTION ENGINE
page 48

THE LAVATORY
page 136

3-SECOND BIOGRAPHIES
JAMES WATT
1736–1819
Scottish inventor who applied his knowledge of thermal energy to warming homes

WILLIAM STRUTT
1756–1830
English mill owner who made buildings safer using iron frames, and warmed them with his new heating methods

FRANZ SAN GALLI
1824–1908
Russian inventor who gained the czar's seal of approval to promote his inventions

EXPERT
Andrew Simms

Our search for better balance with the biosphere begins at home.

APPENDICES

RESOURCES

BOOKS

Blessed Days of Anaesthesia: How Anaesthetics Changed the World
Stephanie Snow
(Oxford University Press, 2008)

Computer: A History of the Information Machine
Martin Campbell-Kelly
(Westview Press, 2013)

ENIAC: The Triumphs and Tragedies of the World's First Computer
Scott McCartney
(Walker, 1999)

Guglielmo Marconi: Building the Wireless Age
Tim Wander
(New Generation Publishing, 2015)

History, Theory, and Practice of the Electric Telegraph
George Bartlett
(Adamant Media Corporation, 2004)

Johannes Gutenberg and the Printing Press
Diana Childress
(Twenty-First Century Books, 2008)

Philo T. Farnsworth: The Father of Television
Donald Godfrey
(The University of Utah Press 2001)

Rebels on the Air: An Alternative History of Radio in America
Jesse Walker
(New York University Press, 2001)

Röntgen Centennial: X-rays in Natural and Life Sciences
Gottfried Landwehr
(World Scientific, 1997)

Smartphones and Beyond: Lesson From the Remarkable Rise and Fall of Symbian
David Wood
(Amazon Media, 2016 Kindle Edition)

Television: An International History of the Formative Years
R. W. Burns
(Institution of Engineering and Technology 1997)

The Alphabet
David Sacks
(Hutchinson, 2003)

The Gutenberg Revolution: How Printing Changed the Course of History
John Man
(Transworld Publishers, 2010)

The Pill: A Biography of the Drug That Changed the World
Bernard Asbell
(Random House, 1995)

The Smartphone: Anatomy of an Industry
Elizabeth Woyke
(The New Press, 2014)

*The Story of Writing: Alphabets,
Hieroglyphs & Pictograms*
Andrew Robinson
(Thames & Hudson, 2007)

The Victorian Internet
Tom Standage
(W&N; New Ed Edition, 1999)

Wheels of Fortune – A Salute to Pioneers
Sir Arthur Du Cros
(Chapman & Hall, 1938)

JOURNAL ARTICLES

"A Brief History of Cardiac Pacing"
Oscar Aquilina
(*Images Paediatr. Cardiol.*, 2006, 27: 17–81.)

WEBSITES

Ancient History Encyclopedia
Jan van der Crabben, Alphabet
ancient.eu/alphabet

Engineers Garage
Samidha Verma, Invention Story of Radio
engineersgarage.com/invention-stories/
radio-history

History of the Microscope
David Bardell, The Invention of the Microscope
www.history-of-the-microscope.org

ThoughtCo.
Mary Bellis, The History of the
Telegraph and Telegraphy
thoughtco.com/the-history-of-the-electric-
telegraph-and-telegraphy-1992542

Mary Bellis, The History of Television
thoughtco.com/television-history-1992530

NASA (National Aeronautics and
Space Administration)
Global Positioning System (GPS)
nasa.gov/directorates/heo/scan/
communications/policy/GPS_History.html

PsPrint, The Invention of the Printing Press
psprint.com/resources/printing-press/

Alan Turing: The Enigma
Andrew Hodges, Alan Turing: The Enigma
www.turing.org.uk

World Wide Web Consortium, W3C
w3.org

World Wide Web Foundation,
The Web Belongs to All of Us
webfoundation.org

NOTES ON CONTRIBUTORS

EDITOR

David Boyle is the author of several books about history, the history of ideas, and the future. His *Authenticity: Brands, Fakes, Spin and the Lust for Real Life* helped put the search for authenticity on the agenda as a social phenomenon, while *Funny Money: In Search of Alternative Cash* launched the "time banks movement" in the UK. His work on the history and future of money has also been covered in books and pamphlets, such as *Why London Needs Its own Currency, Virtual Currencies, The Money Changers: Currency Reform from Aristotle to E-cash, The Little Money Book,* and *Money Matters.* He has stood for Parliament, undertaken an independent review for the UK Treasury and Cabinet Office, and written a number of well-received history books, including *Blondel's Song: The Imprisonment and Ransom of Richard the Lionheart, Toward the Setting Sun: Columbus, Cabot, Vespucci and the Race for America, Alan Turing: Unlocking the Enigma,* and *Before Enigma: The Room 40 Codebreakers of the First World War.*

CONTRIBUTORS

Judith Hodge has worked as a writer and editor in the UK and New Zealand. She is the author of several books and articles on science, health, and green consumerism, including the *New Zealand Green Guide.* After retraining as a public health nutritionist, she is currently coeditor of *Nutrition Exchange,* a publication about nutrition in the developing world.

Diana Rawlinson was the project administrator for both phases 1 and 2 of the DANTE project for Lexicography Masterclass Ltd., creating the *New English–Irish Dictionary.* In 2009–13, she was the legal entity appointed representative for Lexical Computing Ltd on EU Lifelong Learning projects. In 2016, she was a contributor to *The Oxford Handbook of Lexicography.* She now works as a freelance writer.

Andrew Simms is a political economist, environmentalist, and cofounder of the New Weather Institute. He is a research associate at the Centre for Global Political Economy at the University of Sussex and a fellow of the New Economics Foundation (NEF), where he was policy director for more than a decade. During that time, he founded the NEF's work program on climate change, energy, and interdependence, instigated their "Great Transition" project, and ran work on local economies, coining the term "clone towns" to describe the homogenization of main streets by chain stores. He wrote the books *Tescopoly* (on Tesco's dominance of the grocery market in the UK) and *Ecological Debt* (on framing the transgression of planetary boundaries), and he coauthored *The New Economics* and *Green New Deal*, devising the concept of "ecological debt day" to illustrate when in the year we begin living beyond our environmental means. He writes often for *The Guardian* and was described by *New Scientist* magazine as "a master at joined-up progressive thinking." His latest book *Cancel the Apocalypse: The New Path to Prosperity* is a manifesto of new economic possibilities. You can follow him on Twitter at @andrewsimms.uk

INDEX